P9-CQF-742

CONSTELLATIONS

The Stars and Stories

CHRIS SASAKI

Illustrations by Joe Boddy

STERLING PUBLISHING CO., INC.
New York

To Brian:
Find what you love.

Library of Congress Cataloging-in-Publication Data Available

Published by Sterling Publishing Co., Inc.
387 Park Avenue South, New York, N.Y. 10016

Distributed in Canada by Sterling Publishing
℅ Canadian Manda Group, One Atlantic Avenue, Suite 105
Toronto, Ontario, Canada M6K 3E7
Distributed in Great Britain and Europe by Chris Lloyd
at Orca Book Services, Stanley House, Fleets Lane, Poole BH15 3AJ, England
Distributed in Australia by Capricorn Link (Australia) Pty. Ltd.
P.O. Box 704, Windsor, NSW 2756 Australia

Maps by Rich Talcott
Book Design by Chris Swirnoff

Printed in China

Sterling ISBN 0-8069-7635-7

CONTENTS

STARGAZING

THE next time you're out under a clear, dark sky, look up. If you've picked a good spot for stargazing, you'll see an amazing sight—a sky full of stars, shining and twinkling like thousands of brilliant jewels. But this spectacle of stars can also be confusing. Try and point out a single star to someone. Chances are, that person will have a hard time knowing exactly which star you're looking at.

It might be easier if you describe patterns of stars. You could say something like, "See that big triangle of bright stars there?" Or, "Do you see those five stars that look like a big letter W?"

When you do that, you're doing exactly what we all do when we look at the stars. We look for patterns—and not just so we can point something out to someone else. We look for patterns because that's what we humans have always done. Whether we're looking for an animal hiding in the forest, or gazing at a cloud that resembles a face, we look for familiar pictures where at first we don't see anything.

That's what people have been doing for thousands of years as they have looked up at the sky. In fact, thousands of years ago, people did that a lot because the sky was a much more important part of their lives than it is to us today. For one thing, it was easier to see the sky before there were electric lights and tall buildings. For another, people spent many hours looking up at the sky before there were televisions and computer games to keep them indoors and busy at night.

The stars told people when the seasons were changing and helped them know when it was time to plant crops or prepare for spring floods.

People also looked to the sky for answers to bigger questions, like "Who are we?" "Where did we come from?" and "Why is the world the way it is?"

Thousands of years ago, people didn't know what the stars were. The stars were just mysterious lights in the sky. So, they used the sky as a canvas and painted pictures on it—pictures of things that were important to their lives. They looked up at the stars and imagined scenes from their legends and myths—stories that told what they believed about the world.

These star pictures and patterns are called constellations. The first people to name the constellations we know of belonged to many different civilizations. They were Sumerian, Babylonian, Arabic, Indian, Greek, Roman, Chinese, Native North American, and so on.

Nearly 2,000 years ago, a Greek astronomer named Ptolemy wrote a book that named over a thousand stars and 48 constellations. Ptolemy didn't invent these constellations. In fact, some of the constellations Ptolemy included in his book had been known for thousands of years. But Ptolemy was the first to record many of the constellations you find on star maps today.

Over the centuries, others added more constellations, and today you'll find exactly the same number of constellations in the sky as there are keys on a piano. There are 88 constellations, from Andromeda to Vulpecula.

As you'll see, the sky is filled with pictures of animals—real animals like bears, lions, fish, birds, and insects, and imaginary creatures like a flying horse, a sea monster, and a dragon. There's an animal that is half goat and half fish, and a creature that is half human and half horse. There's even a unicorn.

There's a king, a queen, and a princess. There are warriors and heroes. And, there are lots of objects scattered across the sky, too, like arrows, telescopes, microscopes, tools, parts of ships, and crowns.

Do these constellations really look like what they're supposed to look like? If you play "connect-the-dots" with the stars in some constellations, you'll end up with a picture that looks a lot like what it's named after—an animal or person. But, with most constellations, you won't. In fact, most constellations don't look like anything much. They're just groups of stars that have been given names.

There are also patterns of stars that have names even though they aren't really constellations. They're called asterisms. The Big Dipper is an asterism. There's also a baseball diamond, a backwards question mark, triangles and other geometric shapes.

Today, astronomers think of the constellations, not as pictures or patterns of stars, but as areas of the sky. To them, the constellations are like states or provinces on a map. Together, the 88 constellations cover the entire sky.

Thousands of years ago, people noticed that the Sun and the planets didn't wander all over the sky. Instead, they followed the same path through the same 12 constellations, year after year. These are the constellations of the zodiac.

What's the best way to look at the stars?

The best way is to find a comfortable place, far from any lights, where you can see as much of the sky as possible. The farther you are from any lights, the better. Even though you might not see any buildings or street lamps, the glow from cities can light up the sky and make it harder to see dim stars. If there's too much light, it's like trying to see fireflies in the daytime!

It's also a good idea to pick a night when there is no Moon. The Moon is beautiful, but it is so much brighter than the stars that it makes it hard to see them.

You'll want to pick a place where there aren't any trees or mountains to block your view of the sky, too. You might even want to take along a comfortable chair or blanket to lie on.

Also, make sure you're in a safe place. If you're a young person, take a parent or adult with you. Besides being safer, it's more fun to explore the constellations with someone and discuss the star stories you've read about in this book.

Don't worry if you don't have a telescope. There's plenty to see without one. In fact, if you're just starting to learn about the night sky, it's better not to use a telescope anyway. A telescope is great for looking at objects like the Moon and planets, but the best equipment for learning the constellations are your eyes.

Once you're familiar with the constellations and feel like you want to see more, think about buying a pair of binoculars before you buy a telescope. Binoculars let you see lots more than you can see with just your eyes. Then, if you want to see even more, you can start to think about buying a telescope. There are lots of books and magazines that tell you what to look for if you want to buy a pair of binoculars or a telescope. There are even astronomy clubs and people who work at planetariums who can help you.

Reading star maps

When you're out looking at the stars and using the star maps in this book, try this:

Find a piece of red plastic. Tape it over a flashlight. Now you can use the flashlight's red beam to help you read the star maps. If you didn't put the red plastic over the flashlight, the bright light would let you read the map, but make it hard for you to see the stars.

We don't see all the stars all the time. The sky puts on a slightly different show for you every night. We see some stars in winter. Others we see on summer nights. We also see different stars, depending on where we are on Earth. People who live in the Northern Hemisphere see different stars than people who live in the Southern Hemisphere.

Six star maps in this book show you what the night sky looks like at different times of the year, if you live in the Northern Hemisphere. Four maps show you the sky at different times of the year, for stargazers in the Southern Hemisphere.

Here's how to use these maps. Once you've found your stargazing spot, find the North Star in the constellation Ursa Minor. (You'll find out how to do this when you read about the constellation Ursa Major.) Now, turn so you're facing south, directly away from the North Star. Hold the star map right side up in your hands.

Then, lift the star map directly over your head, with the word "North" facing north and the word "South" facing south. The star map should now match the real sky. Stars near the edge of the map should be near the horizon; stars near the center of the map should be high in the sky.

There's a second type of star map. Each constellation in this book has a map of the sky showing the constellation, giving you a closer look at it.

The descriptions of each constellation will tell you where in the sky to look for it—and when. If the constellation is one that is best seen in the Southern Hemisphere, the description will say it is a southern constellation. The descriptions will tell you when during the year to look, but remember that the seasons are reversed in the Southern Hemisphere. When it's summer in the Northern Hemisphere, it's winter in the Southern.

Constellations that are near the north and south celestial poles—the points in the sky directly above the Earth's North and South Poles—are visible on most nights of the year. So, in that case, the description of the constellation won't tell you to look in a certain season.

When I'm looking at the constellations, what am I really looking at?

The stars we see in the night sky are actually huge balls of hot gas, much larger than the Earth and the other planets. We see thousands of stars in the night sky, but there are actually billions and billions of stars in the entire universe. Many are smaller than our Sun, and many are much larger. Some stars are hundreds of times bigger than the Sun!

The stars aren't all the same temperature either. They're all very, very hot, but some are a little cooler than the Sun. The light of these cooler stars is slightly redder than the Sun. Some stars are a little hotter than the Sun and shine a little bluer. After you've spent enough time studying the stars, you'll be able to see this color difference for yourself. Imagine being able to look up at the sky and tell that a star is cooler than our Sun!

The stars are very, very far away. How far?

The fastest spaceships ever built are called *Voyager I* and *II*. They travel at about 35,000 miles per hour. It took months for the *Voyagers* to travel from Earth to Jupiter. And, it took years for them to get to Pluto, the farthest planet from the Sun. If *Voyager I* and *II* were heading toward the closest star, even at 35,000 miles per hour, it would still take them thousands of years to get there! And that's the star that's closest to us!

More than stars ...

Stars aren't the only objects in space that you can see when you're looking at the night sky. There are also huge clouds of gas and dust, much larger than our solar system, called nebulas. We can see a few of these objects without binoculars or telescopes. One of the easiest to see is in Orion. There are more that are visible with binoculars or telescopes.

On any dark, moonless night, far from city lights, you'll also see streaks of light flashing across the sky. These are meteors. Some people call them shooting stars. On most nights, you'll be able to see a few meteors every hour, crisscrossing the sky.

On certain nights of the year, you can see many more meteors than usual. In fact, on some nights, you can see a meteor every minute or so. Watch carefully and you'll notice that the meteors on these nights seem to come from the direction of a single constellation. They may streak across the sky in different directions, but they all seem to radiate from the same point. When you see this, you're seeing a meteor shower.

Different meteor showers come from different constellations. Astronomers name meteor showers after the constellation that the meteors appear to radiate from. The Perseid meteor shower radiates from the constellation Perseus, the Lyrids from Lyra, the Geminids from Gemini, and so on.

Whether it's a star, a meteor, the Moon, planets, or the Northern Lights, there's always something to see in the night sky. The best way to enjoy these wonders night after night is to first learn the constellations. Then, you can make your way around the sky the same way you make your way around your neighborhood. Along the way, the constellations will entertain you with stories and tell you a lot of things you want to know about the sky.

THE CONSTELLATIONS

ANDROMEDA: The Princess

ANDROMEDA, the Princess, lies in the autumn night sky. It's a fairly large constellation, shaped like a long V of stars. The easiest way to find it is to first find the Great Square, in Pegasus, the Flying Horse. The long V of Andromeda begins at one of the corners of the Great Square. In fact, Andromeda and Pegasus share both a star and roles in one of the best stories of the night sky.

According to Greek mythology, Andromeda was a beautiful princess of Ethiopia, the daughter of Cepheus, the King, and Cassiopeia, the Queen. Think you and your parents don't get along? Listen to what Andromeda had to put up with. On star maps, we see Andromeda chained to rocks at the edge of the sea. She's there because her mother angered the sea god, Poseidon, who unleashed a flood and a sea monster on the king and queen. In order to save his kingdom, Cepheus was forced to offer his daughter Andromeda as a sacrifice.

But don't worry. There's a happy ending for Andromeda, thanks to Perseus, the Hero.

We'll get to that later.

You'll find more than a great story in this constellation. Andromeda is also home to the farthest thing you can see without binoculars or a telescope. On a clear dark night, find a secluded spot far from any city lights, trees, or tall buildings. Once you've found your spot, turn off your flashlight, too. Let your eyes get used to the dark—it'll take 10 or 15 minutes. Now, look for Andromeda.When you've found the constellation, look for a very, very faint and fuzzy patch of light—almost like a tiny cloud. It will be between Andromeda's right wrist and right hip.

That faint patch of light is M31, or the Andromeda Galaxy. Galaxies are huge collections of millions or billions of stars. Some are shaped like a spiral, like the Andromeda Galaxy. Some are shaped like footballs or disks. The Andromeda Galaxy is over 2 million light years away. That means a spaceship traveling at the speed of light would take more than 2 million years to get there from Earth.

Andromeda

ANTLIA: The Air Pump

ANTLIA is a small constellation, found in the spring skies of the Southern Hemisphere. It doesn't have any bright stars. As with many smaller constellations, there isn't really much to see when you look at Antlia.

Many of the constellations in the southern skies are a lot smaller than the ones in the northern skies. Also, they didn't appear on very early star maps. For lots of reasons, the patterns of stars seen by ancient people in the Southern Hemisphere never made it on to those maps of the sky—probably because the people who drew the maps lived in Europe and other northern countries.

But when Europeans started exploring the Southern Hemisphere a few hundred years ago, they started inventing constellations and adding them to their maps of the sky.

Antlia is one of 17 constellations that were invented by a French astronomer named Nicolas Louis de Lacaille about 250 years ago. He sailed to South Africa, set up a small observatory in Cape Town, and spent a year mapping 10,000 stars.

When he wasn't mapping stars, he was creating new constellations. Lacaille named his new star patterns after tools and instruments used in science and art, including the air pump, which had just been invented.

So, Antlia and many other southern constellations may not relate exciting myths and stories. But they still tell us about the things that were important to people at the time.

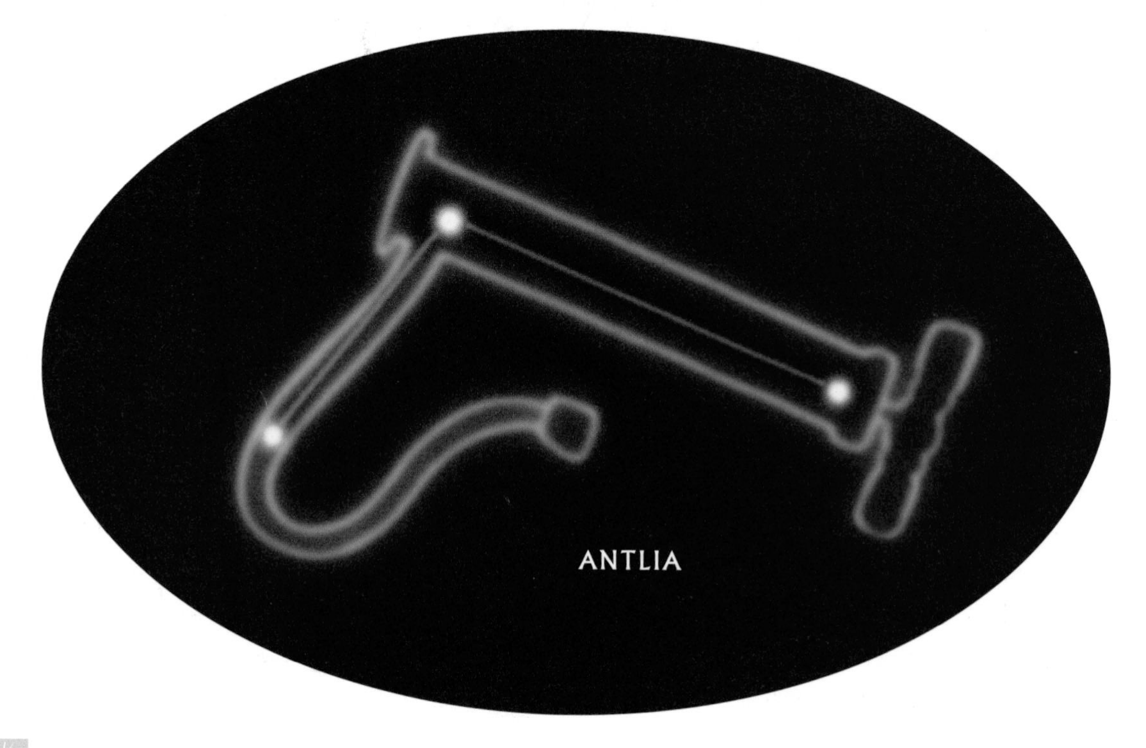

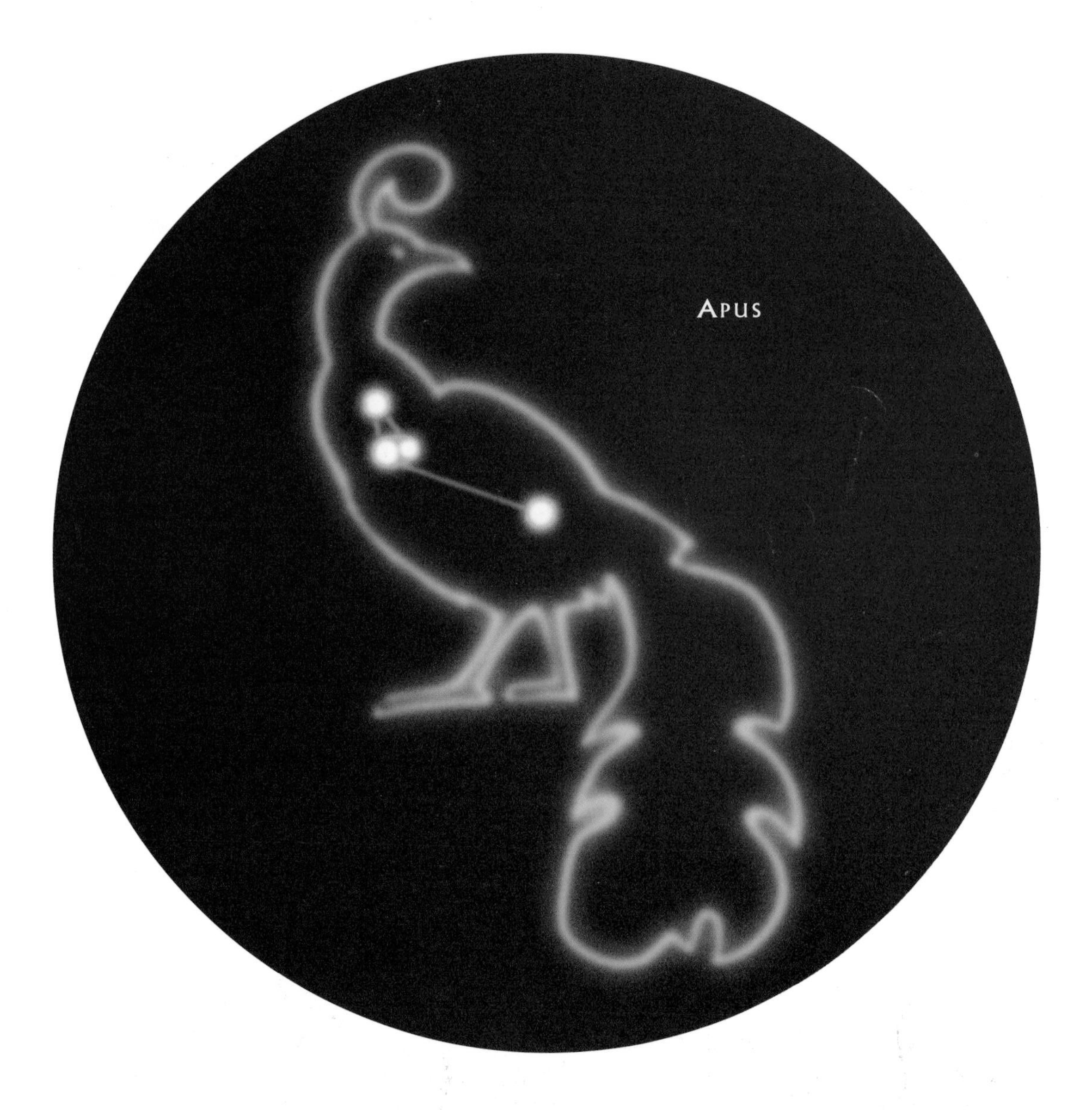

APUS: The Bird of Paradise

Apus is another small southern constellation. Because it's not far from the southern celestial pole, you can see it most nights year-round. There aren't any bright stars or interesting objects in this group of stars.

It is one of 11 constellations invented by the Dutch navigators Pieter Dirkszoon Keyser and Frederick de Houtman, who invented their star patterns while sailing the southern seas over 400 years ago. Keyser and de Houtman named their constellations after exotic animals, and added Apus to the flock of celestial birds already flying through the night sky. Maybe Keyser had birds on his mind because he did a lot of his stargazing from the crow's nest of his ship!

AQUARIUS: The Water Bearer

AQUARIUS is the tenth largest constellation. On star maps, Aquarius is a young man carrying water in a jar. You'll find him in late summer-early autumn skies. The Water Bearer is a constellation of the zodiac that lies between Pisces and Capricornus. This group of stars may not look like someone carrying water, but it does include a Y-shaped group of stars called the "Water Jar."

Aquarius has appeared on star maps for a long time. It was linked with the Sumerian myth of a flood that covered the entire earth. According to the Egyptians, Aquarius emptied his bucket of water into the Nile River, which is why it flooded every year. That wasn't a bad thing, because the flood led the way to a new season of planting. That's why, to the Egyptians, Aquarius was a symbol of good fortune.

There may not be any water falling from the sky from Aquarius, but there are meteors. In fact, there are two different meteor showers that radiate from this constellation, the eta- and delta-Aquarids. You can see the eta-Aquarids in early May and the delta-Aquarids in late July.

Meteors are actually tiny particles that fall into the Earth's atmosphere from space. They're traveling very fast and as they hit the air around our planet, they start to glow. When they glow, we see them as streaks of light.

Where do these particles come from? Some of them are actually small pieces of comets. And different meteor showers are made up of particles from different comets. The eta-Aquarids may not be the most spectacular meteor shower you'll see, but it's fun to think that these meteors may be particles of the most famous comet of all—Halley's Comet!

AQUARIUS

AQUILA

AQUILA: THE EAGLE

AQUILA is a large eagle, soaring through the Milky Way that you can see on summer nights.

The brightest star in Aquila is Altair. It is one of three stars in the Summer Triangle, one of the best known and biggest asterisms in the sky. The stars in the other two corners of the triangle are Vega in Lyra and Deneb in Cygnus.

The Greeks and Romans saw Altair as Zeus's eagle, carrying the thunderbolts hurled by the god. But Altair is also part of a story told in Japan and China that is the Romeo and Juliet tale of the night sky.

The star Vega, in the constellation Lyra, is the Weaving Princess and the daughter of the Sun god. The star Altair is the cowherd who tends the Sun god's cattle. In this story, the two fall in love and marry. And like many newlyweds, they can think of nothing but their love for each other.

The love between the cowherd and the princess grows stronger and stronger with each day, until the lovesick couple begin to neglect their duties. The weaving sits unfinished and the Sun god's cattle roam the fields. Of course, the Sun god is not happy with his daughter and new son-in-law. But try as he might, nothing will get the young couple to pay attention to anything but each other.

Finally, the Sun god's patience runs out and he punishes the two. He banishes them to either side of the river of the Milky Way, and only allows them to see each other one day each year. On the seventh day of the seventh month, magpies fly into the sky and form a bridge across the river, and on this bridge the princess and the cowherd meet. Both are filled with sorrow and joy on that day, and their tears fall to Earth as rain.

ARA: The Altar

ARA is a small constellation, lying in the Milky Way, not far beneath the tail of Scorpius. It can best be seen on summer nights.

To the Greeks, Ara was the altar on which the gods from Mount Olympus swore an oath of allegiance before their great battle against the gods known as the Titans.

ARIES: The Ram

ARIES is a constellation of the zodiac, lying between Taurus and Pisces. You can see it on winter nights. It's a fairly small constellation, without very bright stars, so it's not a well-known pattern of stars, but it does come with an interesting story.

It is the fleece of this ram that makes him famous, because it is what the Greeks called the Golden Fleece. In order to become king, the hero Jason first had to find the Golden Fleece. He set sail in a ship called the *Argo Navis*, with a crew who were called the Argonauts.

The Golden Fleece was to be found in the kingdom of Colchis, ruled by King Aeete. It hung from the branch of an oak tree in a sacred wood, and it was guarded by a huge serpent. Being a practical guy, Jason first asked the king to give him the Golden Fleece. Aeete refused. Luckily for Jason, the king's daughter Medea fell in love with him. She bewitched the serpent, enabling Jason to snatch the prize from the tree.

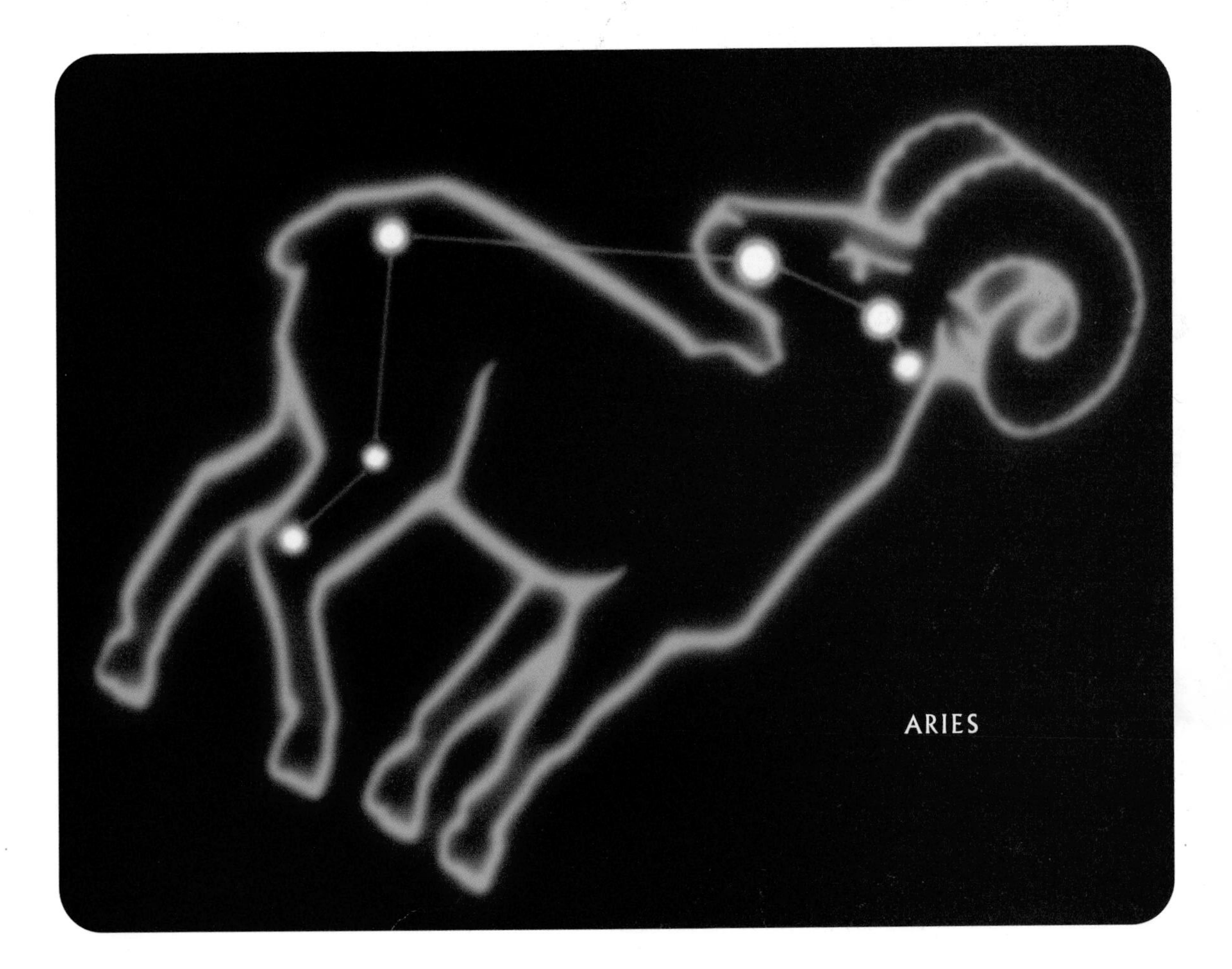

AURIGA: The Charioteer

AURIGA the Charioteer is a medium-sized constellation, racing across the Milky Way in the night skies of winter.

The brightest star in Auriga is named Capella. In fact, it is the sixth brightest star in the night sky. It's also part of an asterism called the "Winter Circle" or "Winter Hexagon," and it can help you find other constellations. Start with Capella. The bright star to the southwest of Capella is Aldebaran in Taurus. To the south of Aldebaran is another bright star—Rigel, in Orion. Now, keep going in a clockwise direction and you'll get to the brightest star in the night sky, Sirius in Canis Major. Next, head back north to Procyon in Canis Minor. Castor and Pollux, the Gemini Twins, are next. Finally, you're back at Capella!

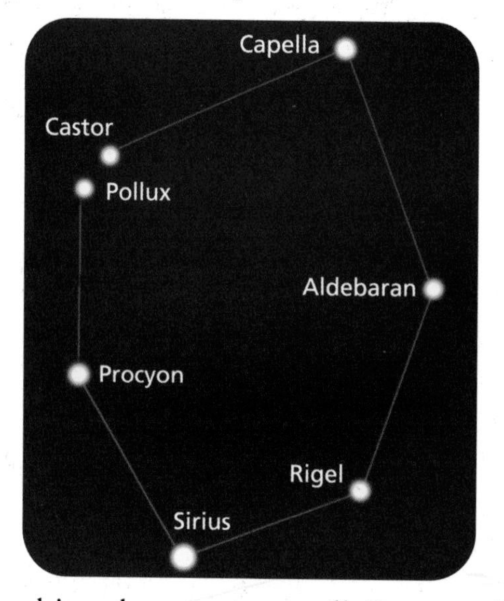

The Salish—people who lived in what we now call Oregon, Washington, and Idaho—tell a story about the stars in the constellations Auriga and Perseus.

In the story, there is a village in the sky where women are preparing the roots of the camas plant for dinner. They cook the roots on hot rocks in a fire pit and soon the smell of the meal fills the forest. It's not long before a skunk wanders into the village, attracted by the fragrance of the camas root and hungry for a meal.

When the women see the skunk waddling toward the fire, they turn to run. But they aren't about to waste all that hard work they did to prepare dinner, so they gather around the fire to keep the skunk away from their food. The women stand with their hands on their hips, determined not to lose their meal. The skunk faces them, determined to have the meal for himself. And so they stand, staring at each other. In fact, they're all so stubborn, they still haven't given up. You can see them every night, in the stars of Auriga and Perseus.

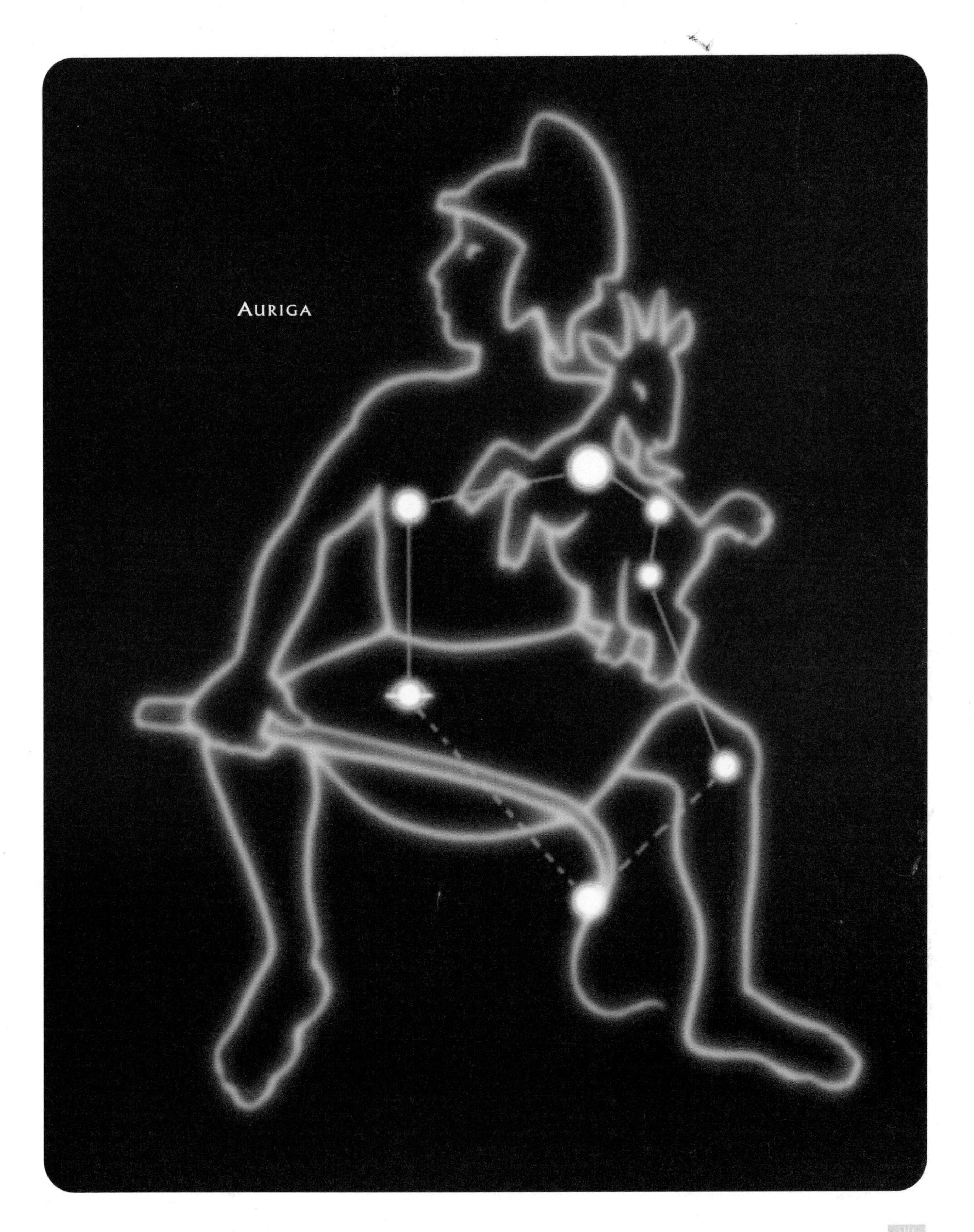
AURIGA

BOÖTES: The Bear Driver or Herdsman

BOÖTES is a spring constellation. It may not look a lot like a bear driver or herdsman, but it's still one of the best pictures in the night sky. The bright stars of Boötes make a shape that looks like a kite, or an upside-down necktie, or an ice cream cone. Some people even think the nearby constellation Corona Borealis is a scoop of ice cream that fell off the cone.

The brightest star in Boötes is also the fourth brightest star in the sky. It's called Arcturus. To find Arcturus, find the Big Dipper. Then, draw a curving or arcing line along the handle of the dipper. That line will go right to Arcturus. You can remember how to find Arcturus by remembering the words "Arc to Arcturus."

Boötes is also known as the Bear Watcher, Bear Guard, or Hunter of the Bear. The bear, of course, is Ursa Major. And Boötes' hunting dogs are the dogs of the constellation Canes Venatici.

The Micmac people live on the east coast of Canada and tell this story of the stars of Boötes and the Big Dipper.

One spring, a bear awoke from its hibernation and emerged from its den. As it was waking and stretching from the long winter, a hunter spotted the bear. This hunter wasn't any ordinary hunter—he was a chickadee. The chickadee quickly fetched a pot for cooking the bear, and called for his hunting companions to join him. Robin brought his bow and arrow. Moose Bird, Pigeon, Blue Jay, Horned Owl, and Saw-whet Owl followed, and the hunt was on.

The birds chased the bear for months. Saw-whet tired first and fell behind. Blue Jay and Pigeon also gave up. Finally, autumn came, and after chasing the bear for almost a year, the hunters caught up to their quarry. The bear was exhausted, so he turned and faced his hunters.

Robin let fly an arrow, which found its mark. The bear's blood sprayed Robin's breast feathers, as well as the leaves of the forest, turning them all to red. The bear eventually died from his wound, but his spirit lived on in another bear. This bear emerged from his den in the spring, to be hunted by the same hunters. Just as the seasons follow each other year after year, the same hunt begins with the coming of every spring.

If you look up in the sky, you'll see the hunters chasing the bear around the sky, night after night, all year long. The four stars in the bowl of the Big Dipper are the bear, and the stars of the dipper's handle and Boötes are the hunters.

The Quadrantid meteor shower comes from Boötes around January 3. The shower isn't named after Boötes. Instead, it's named after the constellation Quadrans Muralis, which disappeared from star maps a long time ago.

Boötes

CAELUM: The Sculptor's Chisel

CAELUM is a very small constellation, with no bright stars, that represents a chisel that a mason would use to carve stone.

It's one of the constellations invented by de Lacaille in the 1700s, while he was observing in the Southern Hemisphere.

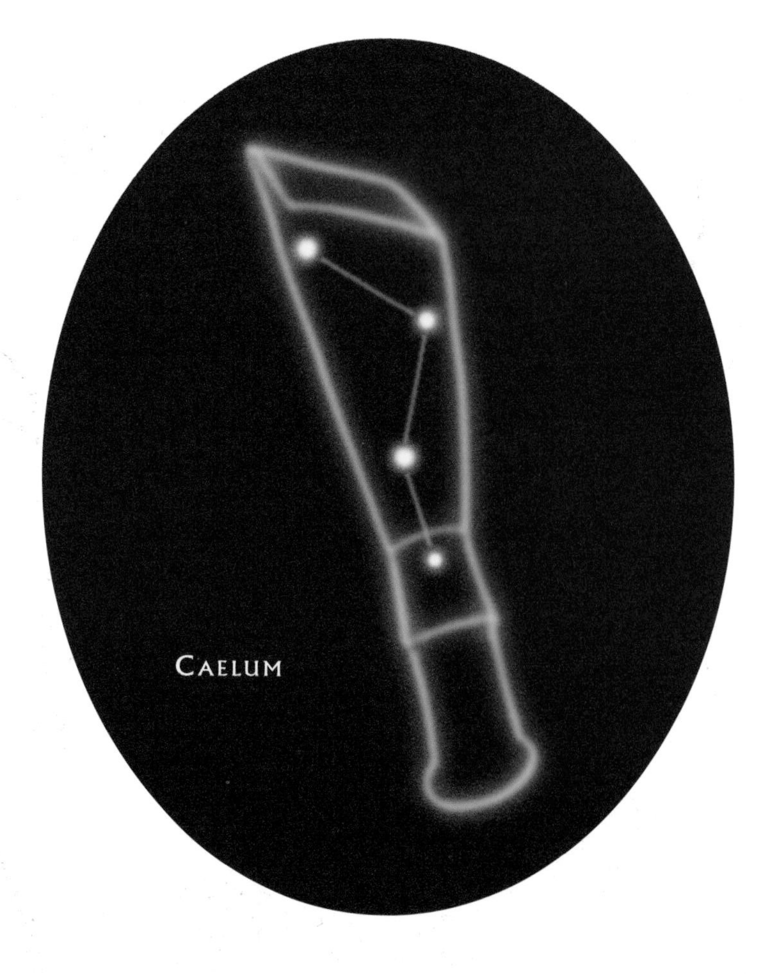

CAMELOPARDALIS: The Giraffe

You wouldn't expect a giraffe to live so close to the North Star, but there it is. Even though it is a large constellation, Camelopardalis is not well known. Its stars pale in comparison to the bright stars in the constellations that surround it—Ursa Major, Cassiopeia, Perseus, and others.

There are no ancient myths about this constellation, because it wasn't invented until 1624. It was first described by a Dutch astronomer named Petrus Plancius.

He called it Camelopardalis, because people used to think of a giraffe as a "camel-leopard," an animal with the head of a camel and the spots of a leopard.

There may not be much to see in this constellation, but a star called 32 Camelopardalis is worth mentioning. The North Star hasn't always been the star closest to the north celestial pole. Thousands of years ago, 32 Camelopardalis was the closest. For the ancient Norse peoples and the Chinese Tang dynasty, this star pointed the way north.

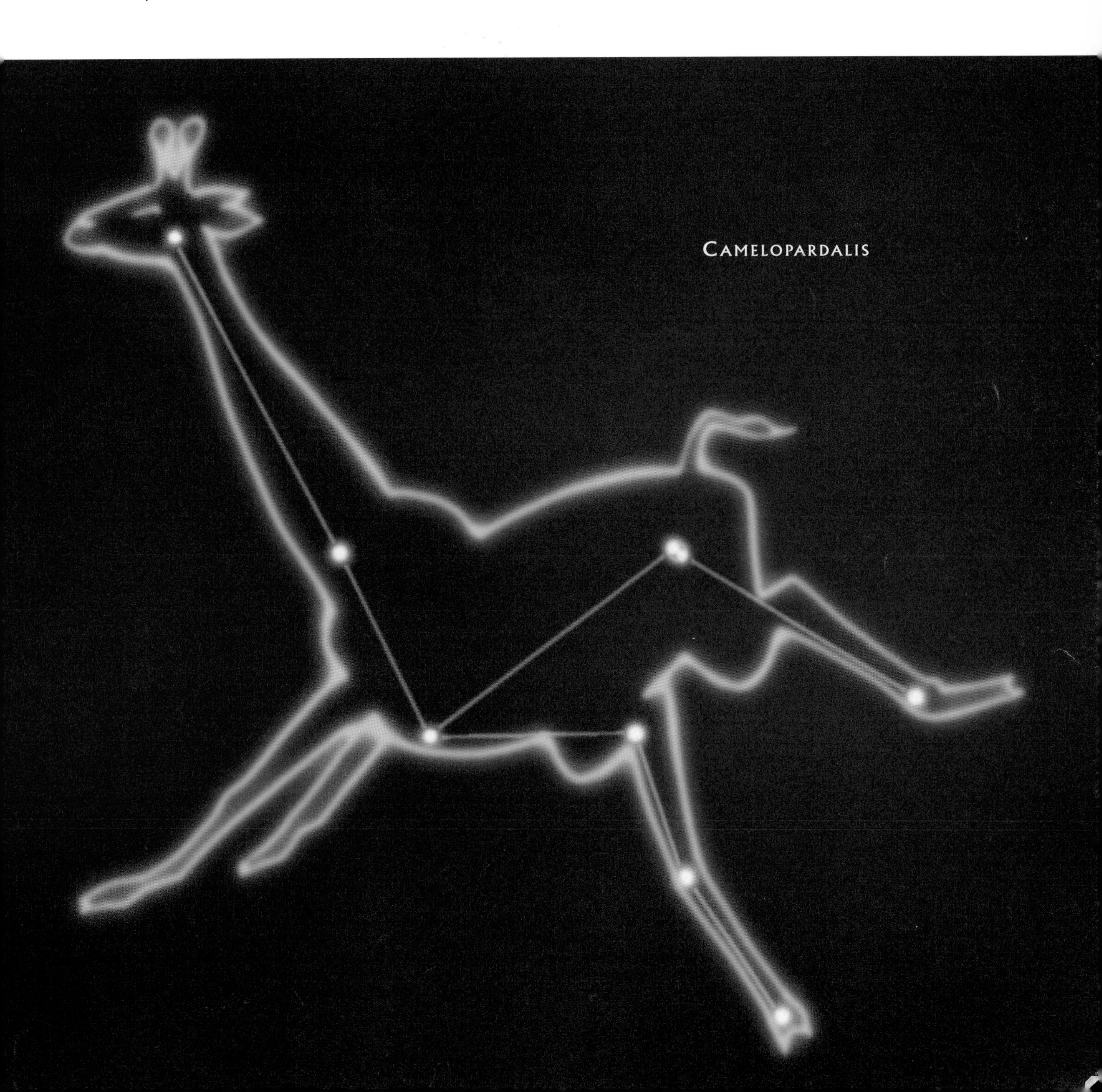

CANCER: The Crab

THE stars of Cancer don't look like a crab. They aren't very bright, either, so this constellation isn't easy to find. But it is a constellation of the zodiac, lying between Gemini and Leo, so you've probably heard of it. We see Cancer in the dark skies of late winter.

According to the Romans, the crab has a small role in the story of one of the Twelve Labors of Hercules. As the great hero fought the many-headed serpent Hydra, an enormous crab joined the battle. Hercules suddenly found himself fighting a tag-team match against two monsters. The crab had been sent by the goddess Juno, who hated Hercules. The crab managed to nip the hero on the heel. But this didn't bother Hercules and he killed both the Hydra and the crab. Juno decided to reward the crab anyway, by placing it in the sky.

When astronomers point their large telescopes at Cancer, they see one of the most interesting objects in the sky. The Crab Nebula is a huge cloud of gas and dust in Cancer. If you think it looks like an explosion, you're right. The nebula is what's left of a star after it exploded. The exploding star was a supernova that was so bright, you could see it during the day. When the Chinese saw it in their skies nearly a thousand years ago, they called it a "guest star." To them, it was a new star that had just arrived in the sky.

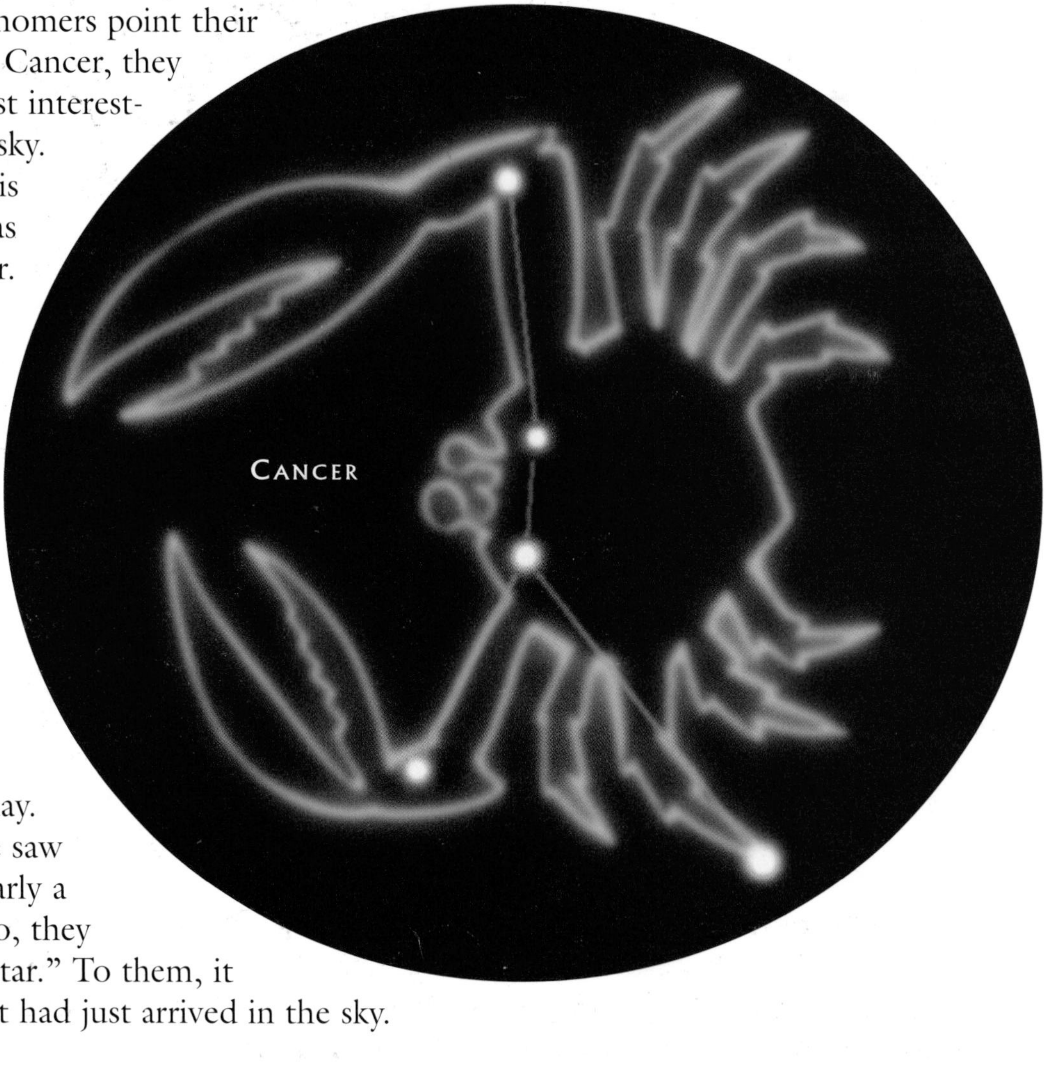

CANCER

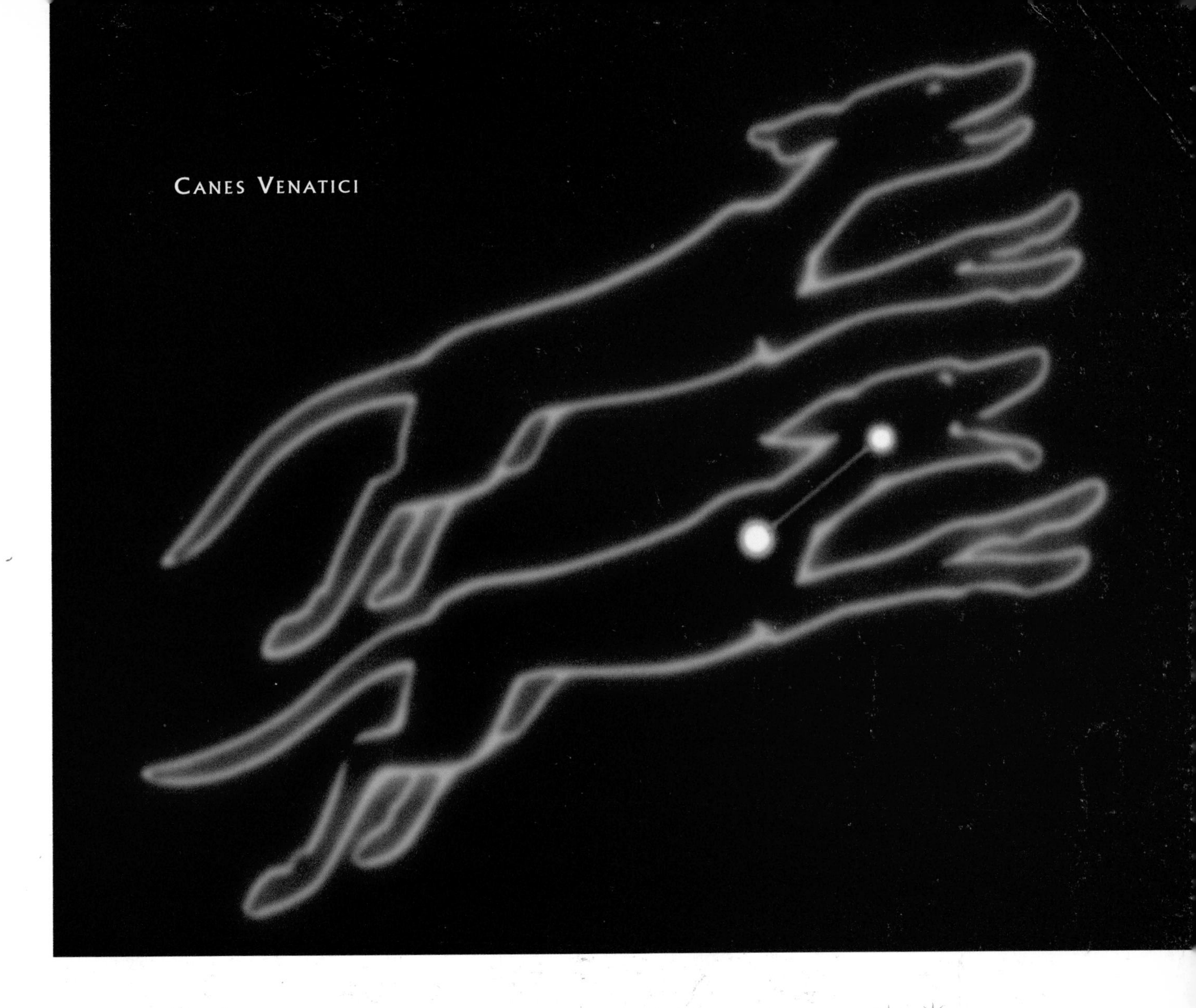

CANES VENATICI: The Hunting Dogs

CANES VENATICI is a medium-sized constellation not far from the north celestial pole, between Boötes and Ursa Major. It doesn't have any bright stars, and the stars that do make up the constellation don't really make a great picture.

Canes Venatici didn't show up on star maps until about 300 years ago. Since then, they have been Boötes' hunting dogs, chasing the great bear, Ursa Major, across the sky.

One of the most spectacular galaxies that we can see is called M51, the Whirlpool Galaxy, in Canes Venatici. It's a spiral galaxy—like ours and the Andromeda Galaxy. From Earth, we can see M51 almost face-on. In photographs taken through large telescopes, we see the galaxy's beautiful spiral arms.

CANIS MAJOR: The Great Dog

CANIS MAJOR is only a medium-sized constellation, but it's one of the best known. That's because is contains the brightest star in the night sky, Sirius, the Dog Star. Because Sirius is so bright, it's very easy to find on winter nights. But in case you need any specific instructions, you can also find this brilliant star by drawing a line through the three stars in the belt of Orion and following that line south and east. It also helps to remember that it's one of the stars of the Winter Hexagon (see page 24).

To the Inuit, people who live in North America's arctic regions, the star Sirius is known as Singuuriq. That means "flickering." According to some of the Inuit, Singuuriq is an old woman. Her house sits beside the path leading from the Earth to the Moon. As the old woman works in her home, the draft from people passing by her door makes her seal-oil lamp flicker and the flame changes color. Whenever a star like Sirius is near the horizon, we see it twinkle in brightness and color. According to this story, the twinkling we see is the flickering of the old woman's lamp.

In another part of the arctic, stargazers see flashes of color for a different reason. Sirius is called Kajuqtuq Tiriganniarlu, which means "Red Fox and White Fox." They say that when we see Sirius change color, we're actually seeing the red and white foxes fighting each other madly, trying to get into a fox hole.

Why is Sirius so bright? If Sirius and the Sun were side by side, you'd see that the Dog Star is actually 25 times brighter than the Sun. But that's not the reason it's the most brilliant star in the night sky. After all, there are lots of stars brighter than the Sun. The reason it's so bright is because Sirius is only 9 light-years away, and there aren't many stars closer to us than that. In fact, out of all the billions of stars in our galaxy, there are only six star systems closer than Sirius. So, the brightest star in Canis Major is also the brightest in the entire sky because it's nearby.

CANIS MINOR: The Little or Lesser Dog

CANIS MINOR is a fairly small constellation, with only a handful of stars that look nothing like a dog. But, like Canis Major, it contains a very bright star. Procyon is the eighth brightest star in the sky. Like Sirius, Procyon is part of the Winter Hexagon.

To the ancient Greeks, Procyon means "before the dog" because it rises before Sirius, the Dog Star. But to the Inuit, Procyon is known as Sikuliaqsiujuittuq, "the one who never goes onto the newly formed sea-ice."

Why doesn't he go onto the sea-ice? According to the Inuit, it's because he's too heavy. And since he can't go onto the sea-ice and fish like the others, he has to steal his catch from the fishermen. Obviously, to the Inuit, Procyon is not a popular guy!

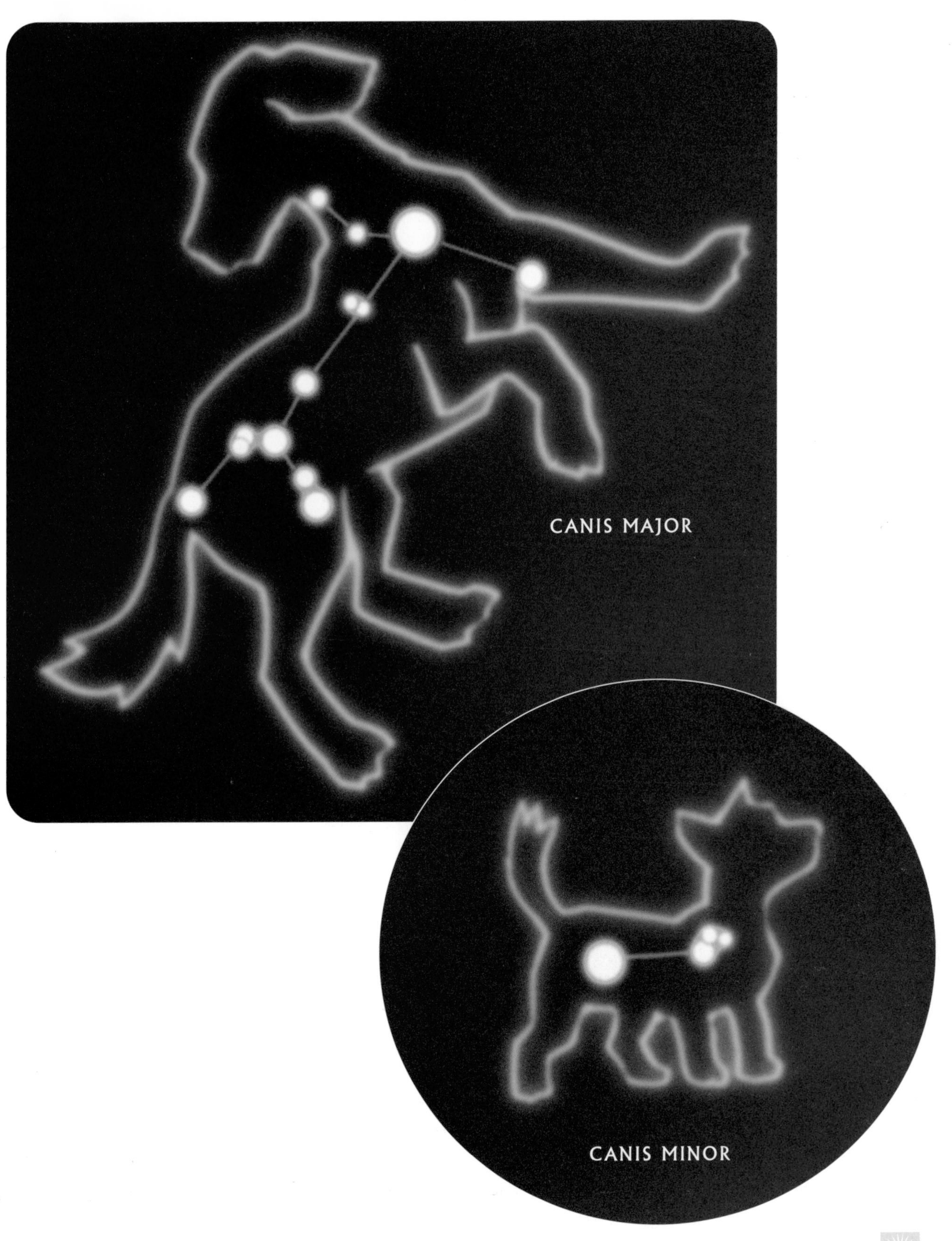
CANIS MAJOR
CANIS MINOR

CAPRICORNUS: The Sea Goat

CAPRICORNUS is the smallest of the constellations of the zodiac, sitting between Sagittarius and Aquarius in late summer skies. Even if you knew what a "sea-goat" was, it would be hard to see such a creature in the stars of Capricornus.

It's easy to overlook, but Capricornus is one of the oldest constellations in the sky.

In Greek legend, Pan was the god of shepherds and flocks. According to one story, he once took the form of a goat and, while trying to escape a monster, jumped in a river and began to transform into a fish. But the river wasn't quite as deep as Pan imagined. Only the part of his body below the water turned into a fish. The upper part of his body, still above water, stayed as a goat. This is the strange creature we see on star maps today.

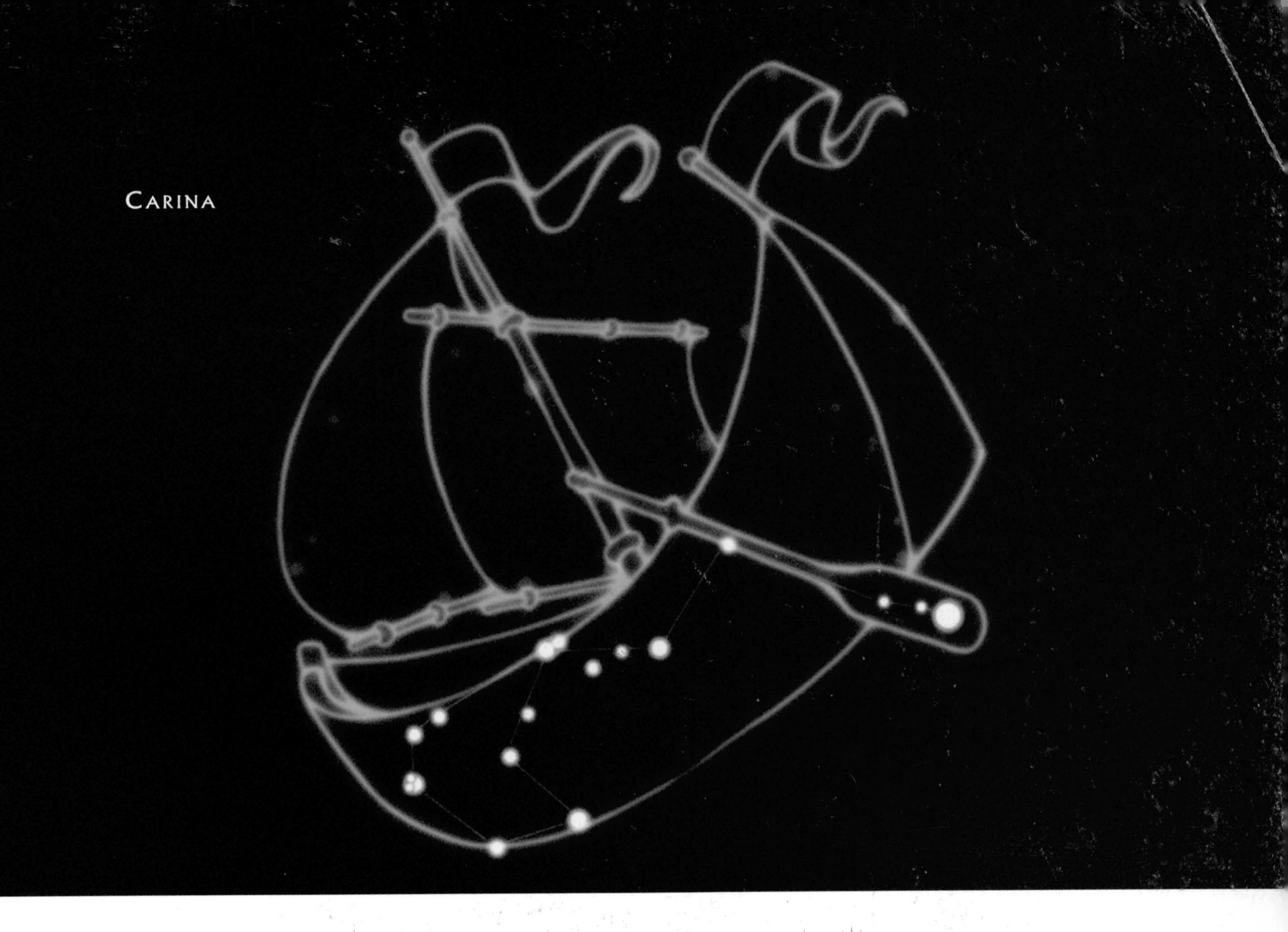

CARINA: The Keel

CARINA is a fairly large constellation, not far from the south celestial pole, sailing along the shores of the southern Milky Way. The brightest star in Carina is Canopus, which is the second brightest star in the night sky. It is high in the sky during southern summers.

Carina used to be part of a very large constellation called Argo Navis. In Greek legend, Jason and the Argonauts sailed the *Argo Navis* on a voyage in search of the Golden Fleece. The crew included the Gemini twins, Castor and Pollux, as well as Hercules. With such a stellar crew, it's no wonder that Jason recovered the Golden Fleece and returned home in triumph. Athena, the goddess who watched over the ship and her crew, placed the ship in the sky where it could sail a sea of stars forever.

When Nicholas Louis de Lacaille mapped the stars of the southern sky 250 years ago, he divided the constellation Argo Navis into three smaller constellations. The stern of the ship became known as Puppis. The sail was called Vela. And the keel became Carina.

CASSIOPEIA: The Queen

CASSIOPEIA, the Queen, sits on her throne in the northern skies. You can imagine that the five bright stars of Cassiopeia are a throne, but to most people this constellation is just a big W in the sky. Since it's not far from the north celestial pole, you can usually find Cassiopeia somewhere in the sky, no matter what the season.

In mythology, Cassiopeia was the Queen of Ethiopia. She was married to King Cepheus and was the mother of Andromeda. Cassiopeia was very vain. She thought she was more beautiful than anyone, even the sea nymphs known as the Nereids. Of course, that didn't go over well with them or with the sea god Poseidon. He sent a sea monster to attack her kingdom and teach her some humility.

Did it work? It doesn't look like it. On some star maps, Cassiopeia still sits on her throne, combing her hair and admiring herself in a mirror.

To the Inuit of the Arctic, the W was a flight of steps cut in the heavenly snows, steps between the sky country and the earth. To natives of the northwest coast, the W was the hide of an elk skin. The elk was so large that the only place to dry it was in the sky. The five stars were holes, where stakes were driven through the hide in order to stretch it.

Over 500 years ago, there were six bright stars in Cassiopeia when a new, very bright star appeared in the constellation. It was a supernova, an exploding star. The supernova was so bright, you could see it in the daytime. Imagine stargazing in the middle of the afternoon!

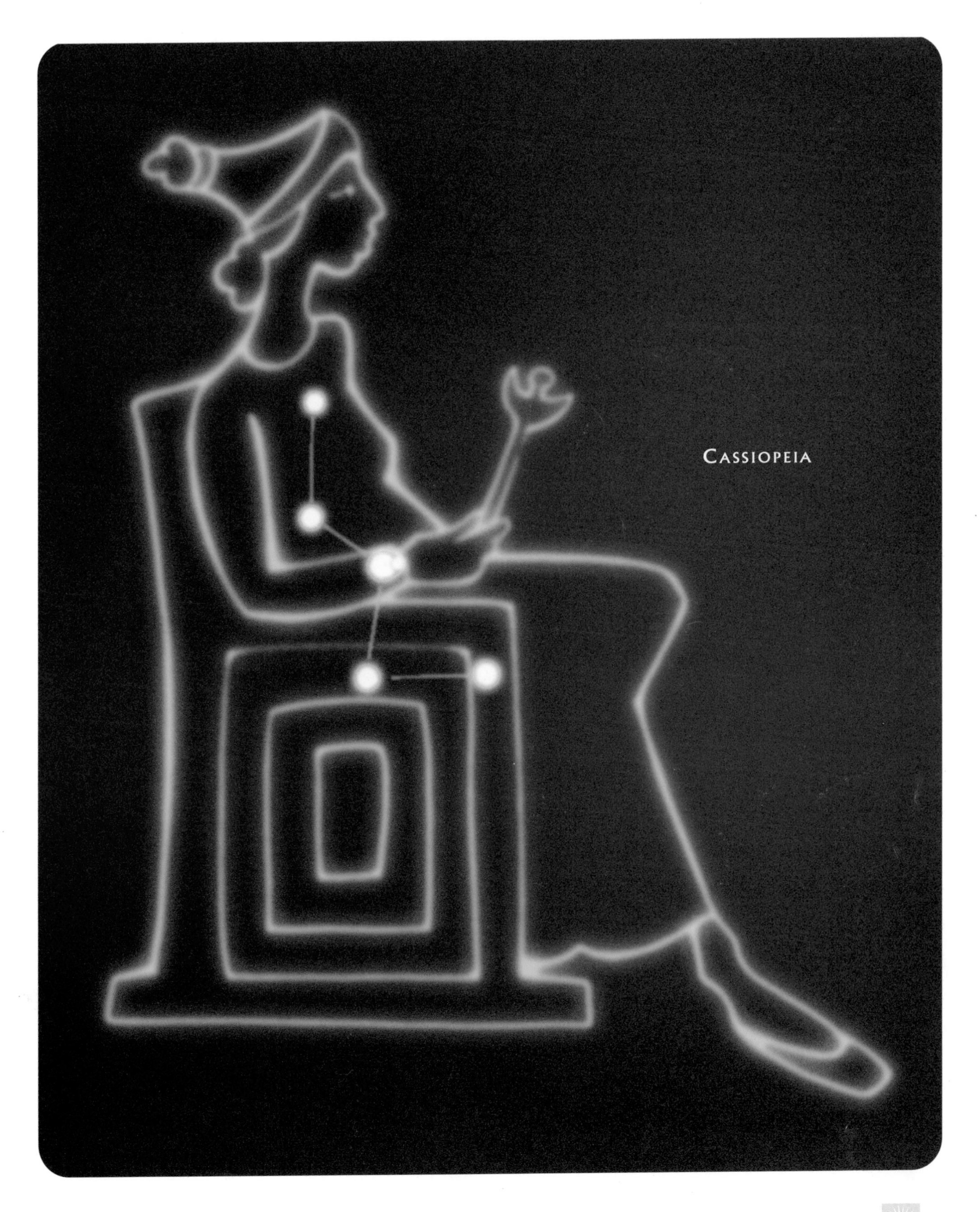
CASSIOPEIA

✶✶✶ CENTAURUS: The Centaur ✶✶✶

CENTAURUS is the ninth largest constellation, and is high in the sky during southern autumn months. It gallops along the southern Milky Way, and contains one of the brightest and most famous stars in the sky, Alpha Centauri.

A centaur is another imaginary beast, half human and half horse. To the Greeks, the constellation Centaurus is Chiron. Most centaurs were wild beasts and were known to get fierce and rowdy every now and then. Chiron was different. He was wise and taught hunting, medicine, and music to the children of the gods.

Alpha Centauri is the brightest star in the constellation and the third brightest in the sky. It's actually a group of three stars. This trio is famous because it is the nearest star system to us—other than the Sun, of course. Alpha Centauri is about 4½ light-years from us. That means that light from this star takes 4½ years to reach us. (If you really want to be accurate, one of the three stars of Alpha Centauri is closer to us than the others. It's called Proxima Centaur.)

A light-year isn't a length of time, it's a distance. It's what astronomers use to measure distances in space. Stars, nebulas, and galaxies are so far apart, it's hard to say how many miles away they are. For example, 4½ light-years is the same as 26,500,000,000,000 miles. It's a lot easier to write 4½ than 26,500,000,000,000!

Centaurus

CEPHEUS

CEPHEUS: The King

CEPHEUS is a house-shaped pattern of stars. Because it's not far from the North Star, you can see it on most nights throughout the year.

Cepheus was the king of Ethiopia, and you can see the rest of his family in the sky, too—Queen Cassiopeia and their daughter Andromeda. But it's not a portrait of a happy family. When Poseidon sent a sea monster to destroy his kingdom, Cepheus was forced to sacrifice his daughter Andromeda by chaining her to the rocky coast.

Delta Cephei, one of the brighter stars in the constellation, is also one of the most famous and important stars in astronomy. Delta Cephei is a variable star, a star that changes in brightness. Astronomers gave variable stars like Delta Cephei a name: Cepheid variables.

About a hundred years ago, an astronomer named Henrietta Leavitt studied Cepheid variables and made an important discovery. She found that you could tell the distance to a Cepheid variable star simply by knowing how long the star took to change in brightness.

Why was this so important? It gave astronomers a new way to measure very long distances in space. With Leavitt's discovery, they could now look for Cepheid variables in other galaxies and measure the distance to those galaxies—something they couldn't do before. With Cepheid variables, astronomers had a much better idea of just how big the universe was.

CETUS: The Sea Monster, The Whale

WHEN the sea god Poseidon sent a sea monster to destroy Cepheus' kingdom, he wasn't fooling around. Cetus is the fourth largest constellation in the sky. It crashes through our autumn skies.

But, as big as the monster was, that didn't stop the great hero Perseus from attacking it and starting a great battle.

One of the stars in Cetus has a wonderful name. It's called Mira, which is Latin for "wonderful." The star is also known as Omicron Ceti. It's a variable star—a star that changes in brightness. It was the first variable star to be discovered. When it is brightest, you can see it with the naked eye, but at its dimmest, it disappears! You need binoculars or a telescope to see it then.

Mira gets brighter and dimmer because the star actually gets bigger and smaller. There are other variable stars that change in brightness for other reasons. Some just shine brighter without changing size. Others seem to change brightness because other dimmer stars pass in front of them and block some of their light from reaching us.

Try watching Mira throughout the year, and see its wonderful disappearing act!

CETUS

CHAMAELEON: The Chameleon

CHAMAELEON is a very small constellation, near the south celestial pole. It's difficult to see because of its size and because its stars aren't very bright. (Unless you think it's because the chameleon is doing a good job of camouflaging itself, just like the real lizard!)

Because it's so close to the south celestial pole, you can see it on any night of the year—as long as you live in the Southern Hemisphere, that is.

Chamaeleon is one of the 11 constellations invented over 400 years ago by two Dutch navigators.

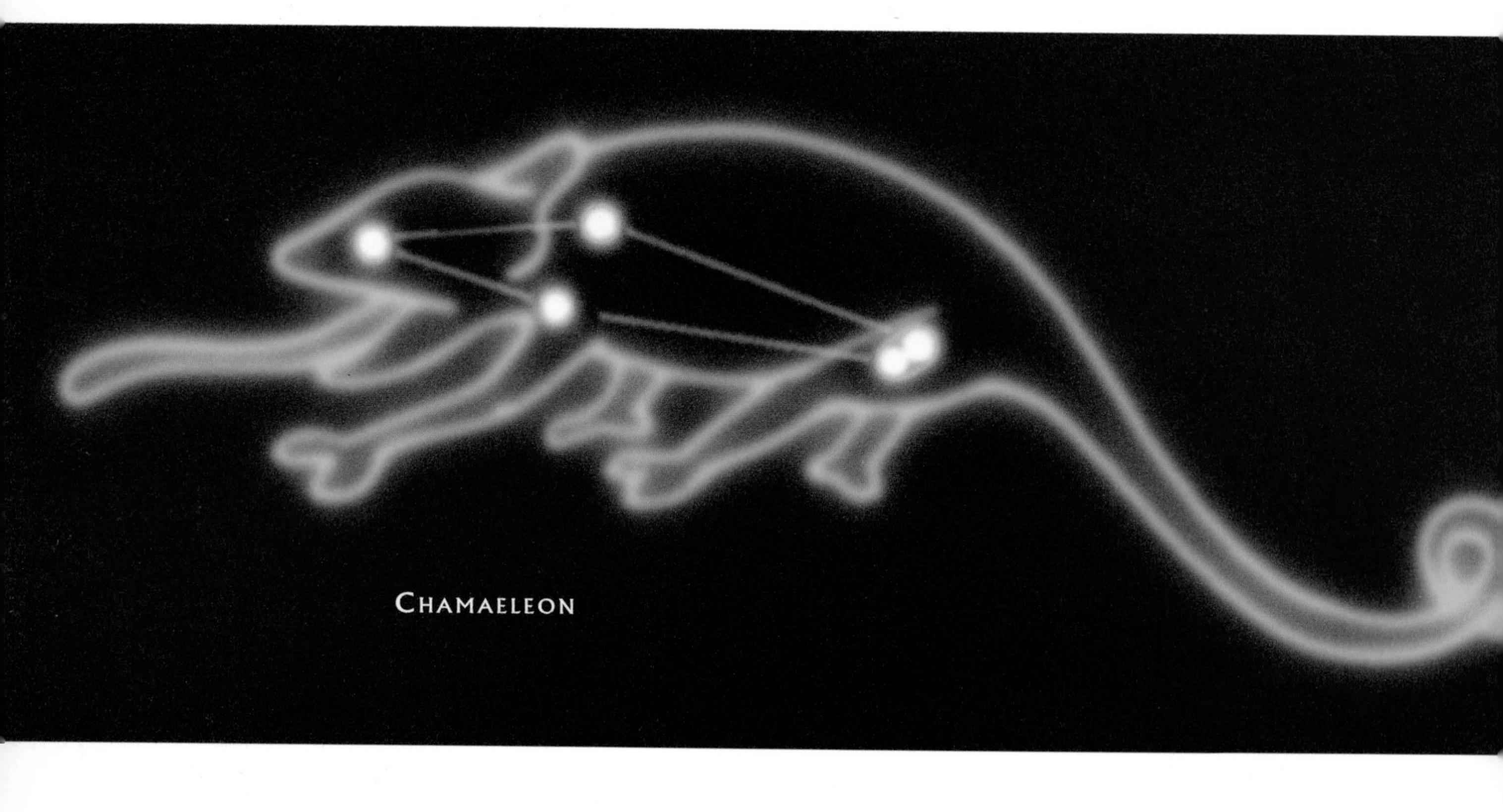

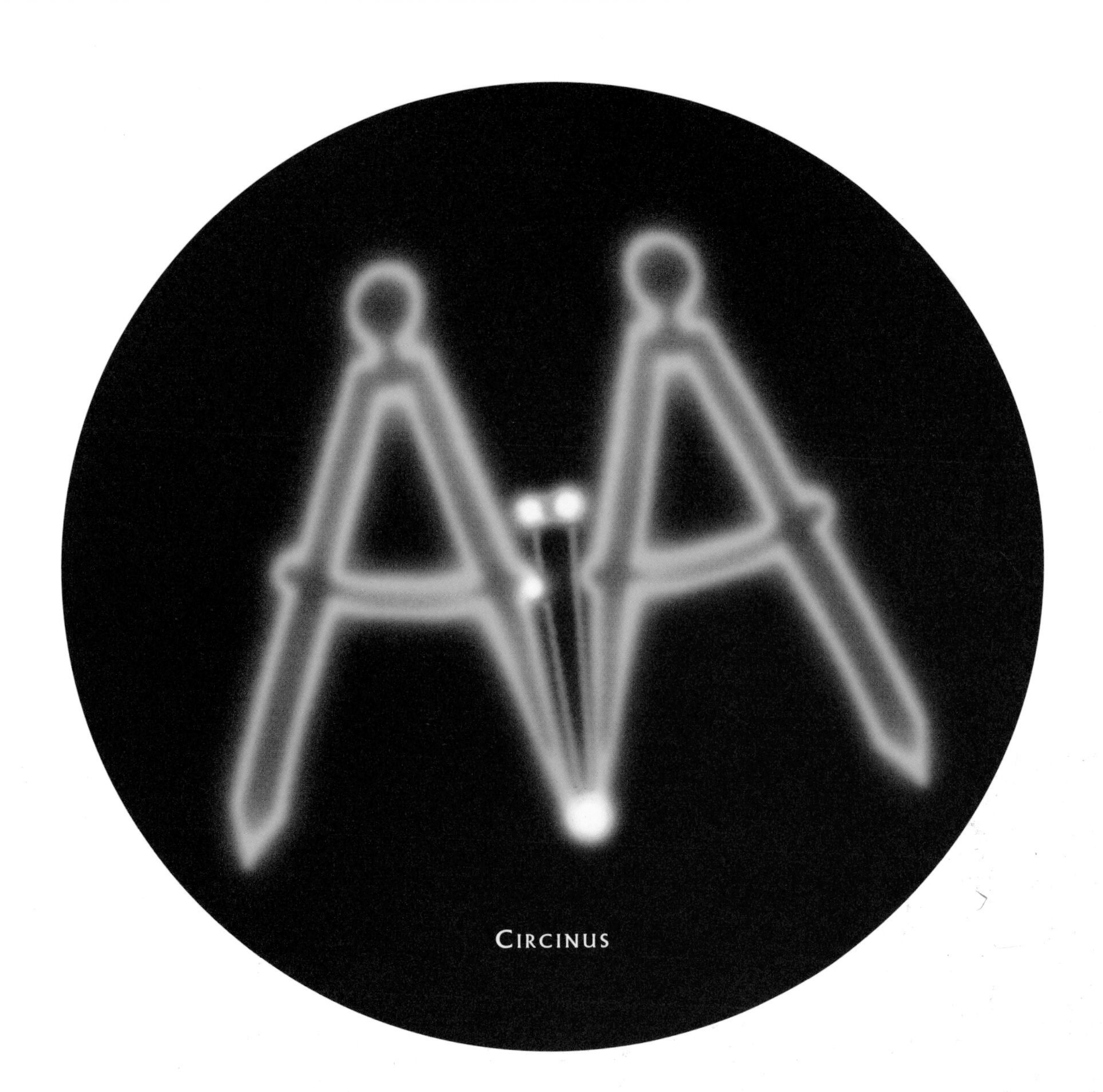

CIRCINUS: The Compasses

CIRCINUS doesn't stand a chance. Not only is it the fourth smallest constellation in the sky, it's right next to Centaurus. The dim stars of Circinus pale in comparison to Alpha and Beta Centauri. If you do look for it, you'll find it lying in the Milky Way. It's best to look during the southern autumn months.

Circinus is a pair of compasses—not the kind that show direction, but the kind that mapmakers and navigators use. This constellation was invented by de Lacaille in the mid-1700s, along with other constellations named after tools and equipment.

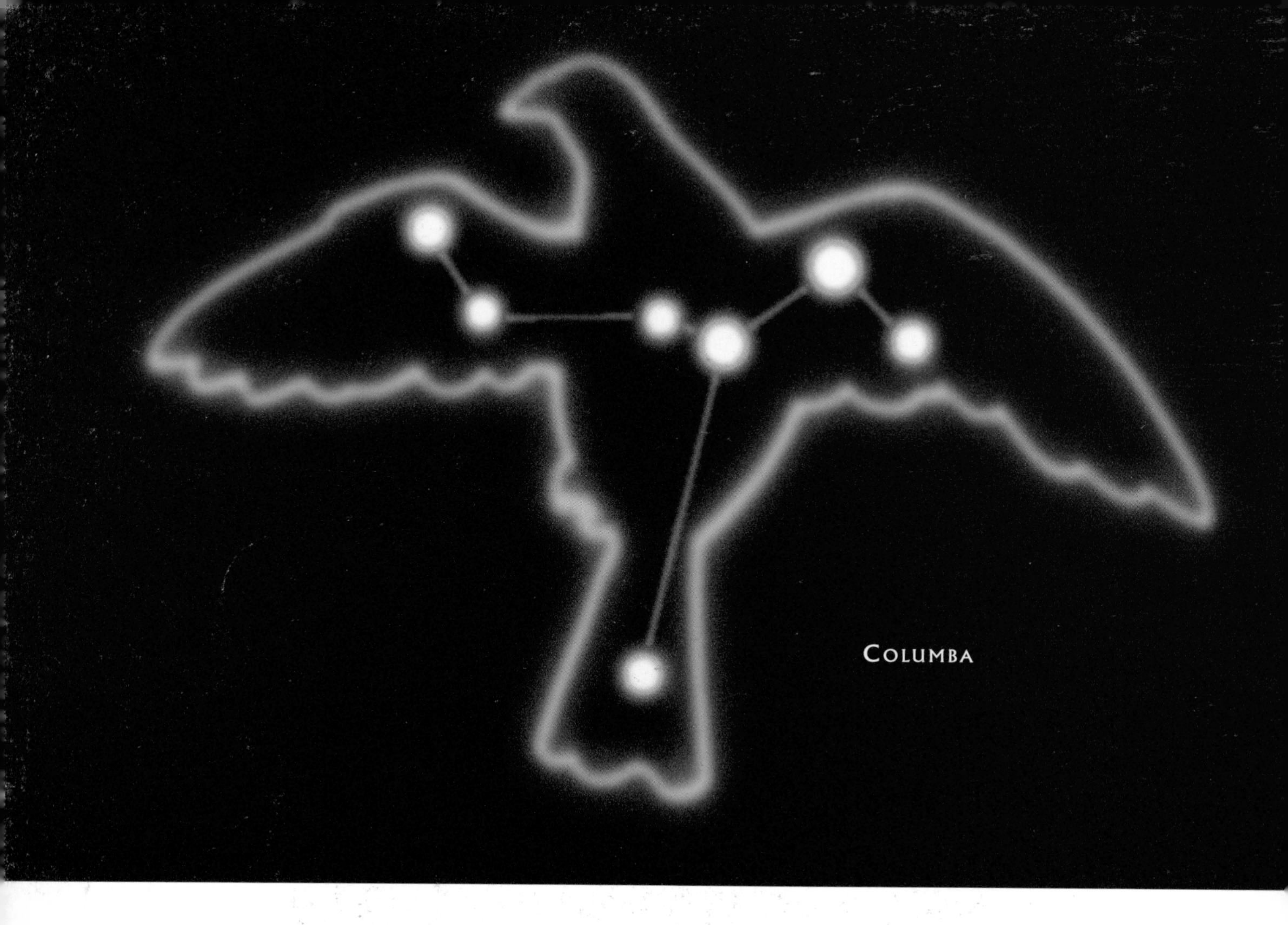

COLUMBA: The Dove

COLUMBA is a small constellation that doesn't really look like a dove. Not only is it small, but all of its stars are outshone by nearby Canopus and Sirius. You'll see it near Canis Major, during the winter.

Columba is the only constellation from the Bible. It is the dove that Noah sent from the ark in search of land. On most star maps, the dove is obviously on its way back to the ark after finding land because it is carrying a branch in its beak. Where's the ark? In this story, the nearby constellations of Carina, Puppis, and Vela make up Noah's ark instead of the ship *Argo Navis.*

If our Milky Way galaxy had a speed limit, there's a star in Columba that would be breaking the law. It's called Mu Columbae and it's speeding through space at 60 miles per second. Two fainter stars are also moving at about the same speed and in the same direction. It may be that these stars began their high-speed chase when a star exploded millions of years ago and sent them flying.

COMA BERENICES: Berenice's Hair

COMA BERENICES is a spring constellation. Without any bright stars, it isn't easy to see—even though it sits between Leo, Virgo, Boötes, and Ursa Major. It doesn't really look like locks of hair, but you could imagine these stars as the tuft of hair on Leo the Lion's tail. In fact, that's how people saw this collection of stars, until they were named after the tresses of a queen.

Coma Berenices represents a real person—or at least her hair. Berenice II was the queen of Egypt over 2,000 years ago, and her husband was King Ptolemy III. When Ptolemy returned safely from battle, Berenice was so relieved that she thanked the gods by cutting off her hair. She placed the hair in a temple, but it quickly disappeared. Some say that when the queen's hair disappeared from the temple, the stars of Coma Berenices appeared in the night sky.

CORONA AUSTRALIS: The Southern Crown

EVEN though Corona Australis is small and has no bright stars, this arc of stars makes a pretty pattern in the summer sky, between Sagittarius and Scorpius. Many people think the stars of Sagittarius make a pattern that looks like a teapot. If so, then the arc of stars of Corona Australis are a slice of lemon, ready for a nice, hot cup of tea.

To the ancient Greeks, the Southern Crown was actually a wreath. And to the native North Americans known as the Shawnee, this group of stars was a circle of star maidens, dancing in the sky.

CORONA BOREALIS: The Northern Crown

CORONA BOREALIS isn't very big and doesn't have any bright stars, but it's still a lovely crescent of stars. It lies between Boötes and Hercules, and is high in the sky in the early summer.

According to one legend, the Northern Crown helped the great warrior Theseus defeat one of his most fearsome enemies.

The Minotaur was a monster with the body of a man and the head of a bull. It lived in a vast palace called the Labyrinth, a palace with so many rooms and hallways that those who entered never found their way out again. Trapped, they became victims of the Minotaur.

The crown belonged to Princess Ariadne of Crete, who fell in love with Theseus. She gave him the crown, and he used it to light his way through the maze of the palace. Once inside, the great hero killed the Minotaur and found his way out of the Labyrinth using a magical thread Ariadne had given him.

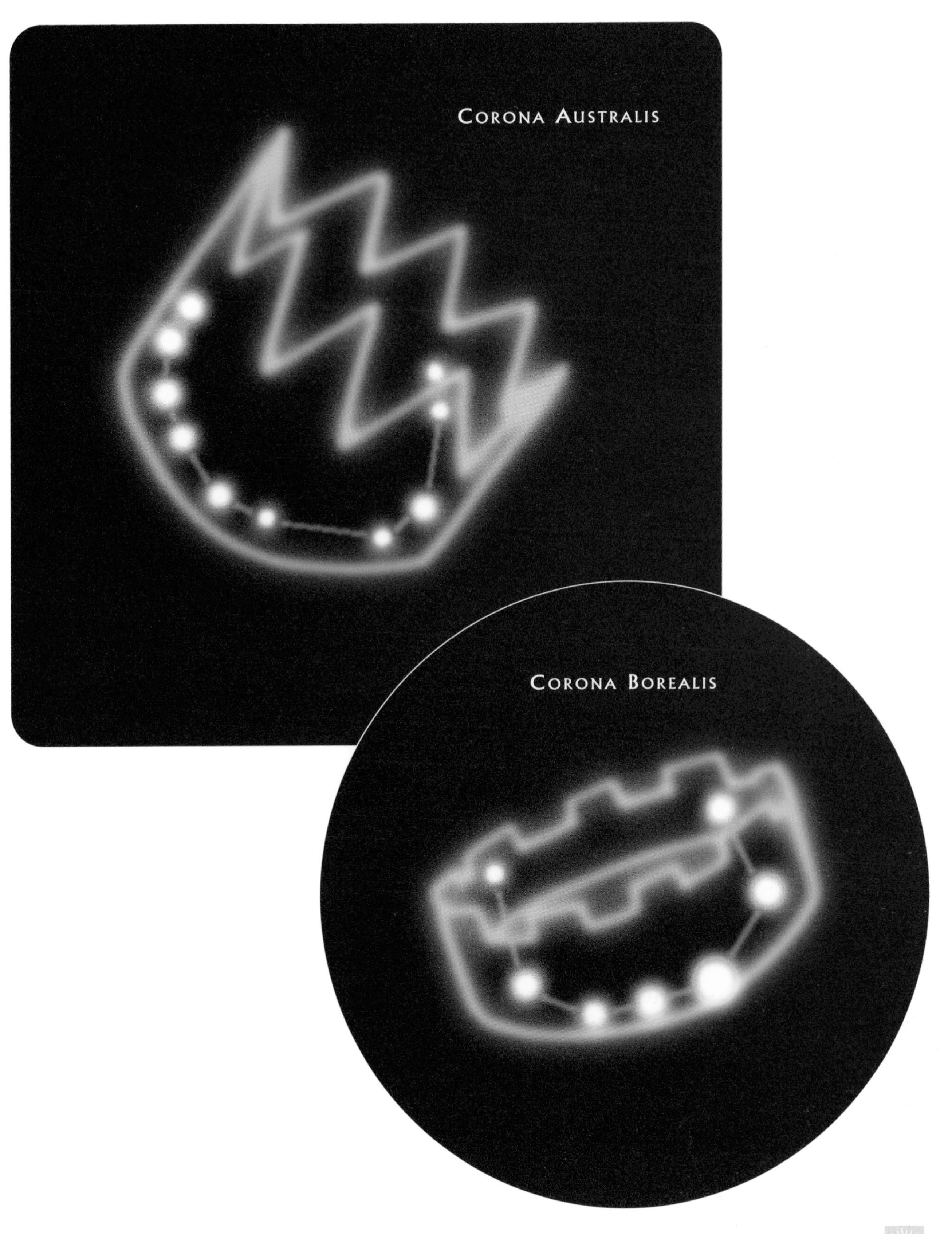
Corona Australis
Corona Borealis

CORVUS: The Crow, The Raven

CORVUS is a very small, springtime constellation. It sits near the much larger Virgo, and near Crater and Hydra.

According to the Greeks, the god Apollo sent the crow to fetch some of water. But the crow was not a very responsible bird. Along the way, he came upon a fig tree and thought that he would like to try some of the figs. They weren't ripe yet, so he waited and waited and waited until they were ready to eat.

When he finally returned to Apollo, the crow made up an excuse for being late. He told the god that Hydra the Water Snake had kept him from completing his task. Well, Apollo knew Corvus was lying, so he punished the crow by putting him in the sky. As if that weren't bad enough, Apollo placed the crow near Crater the Cup. The irresponsible crow can see the cup of water every night, but he's not allowed to drink.

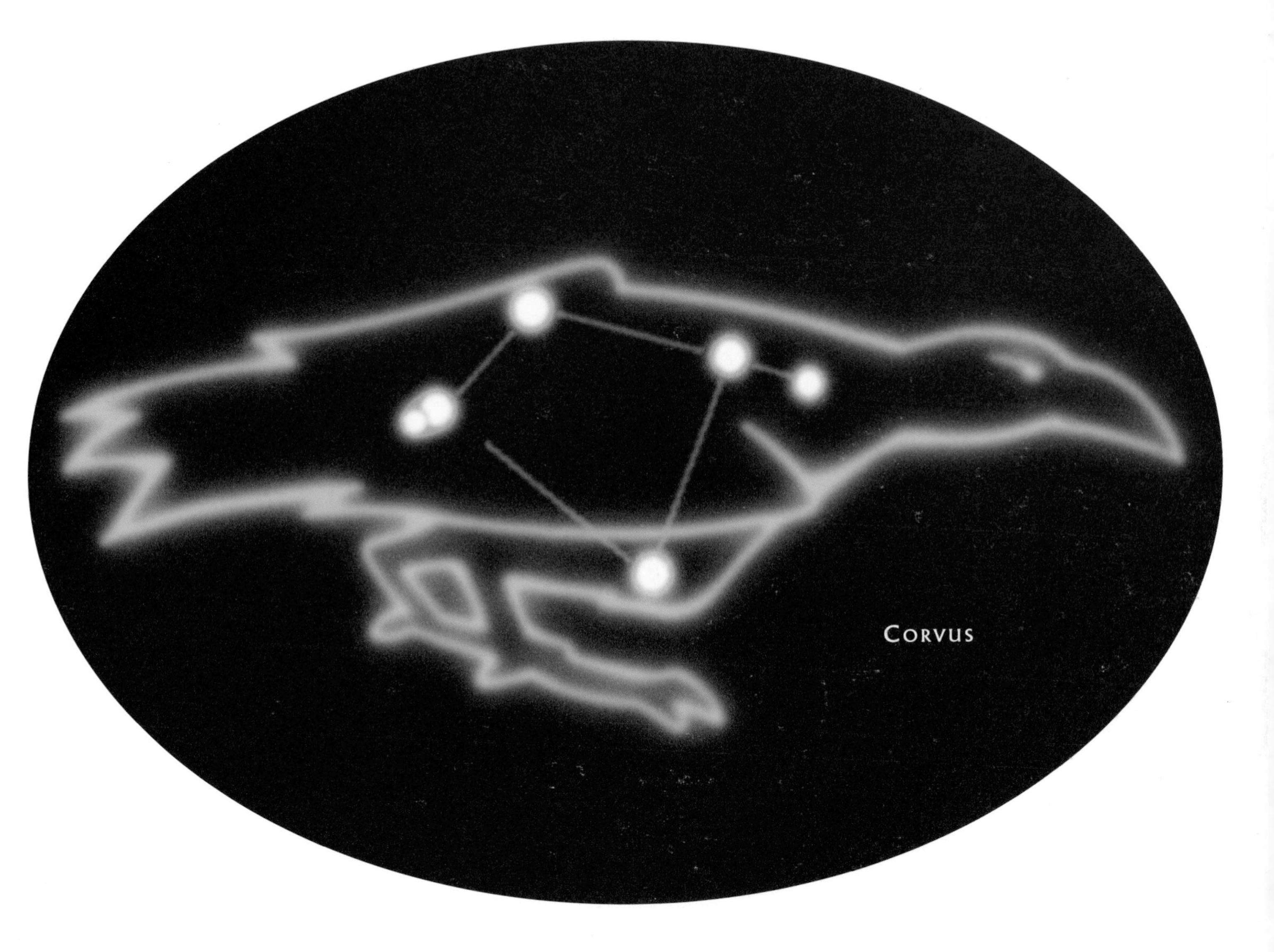
Corvus

CRATER: The Cup

CRATER is a small, springtime constellation. It sits beneath Leo, and near Corvus and Hydra.

According to one legend, Crater is the cup carried by Corvus the Crow. In another, it is a cup of nectar drunk by the Olympian gods.

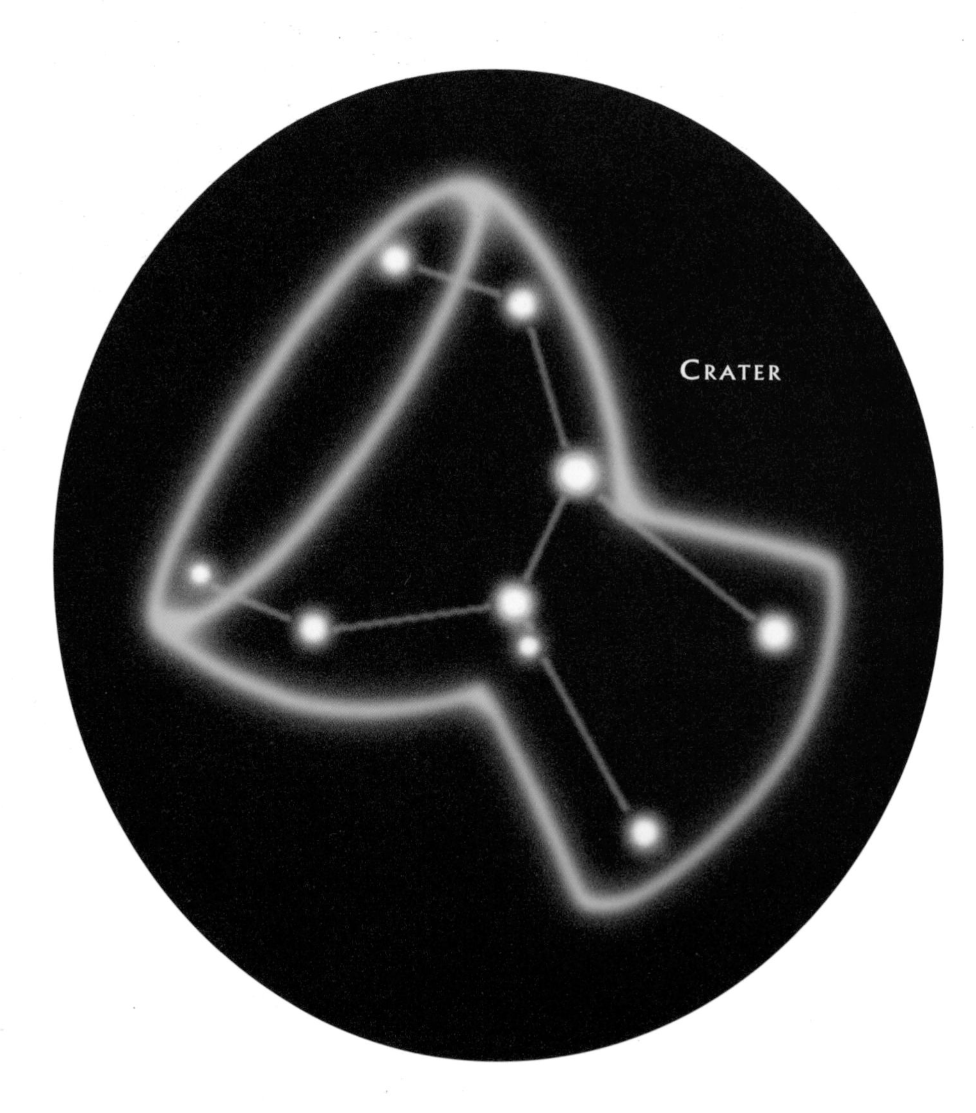

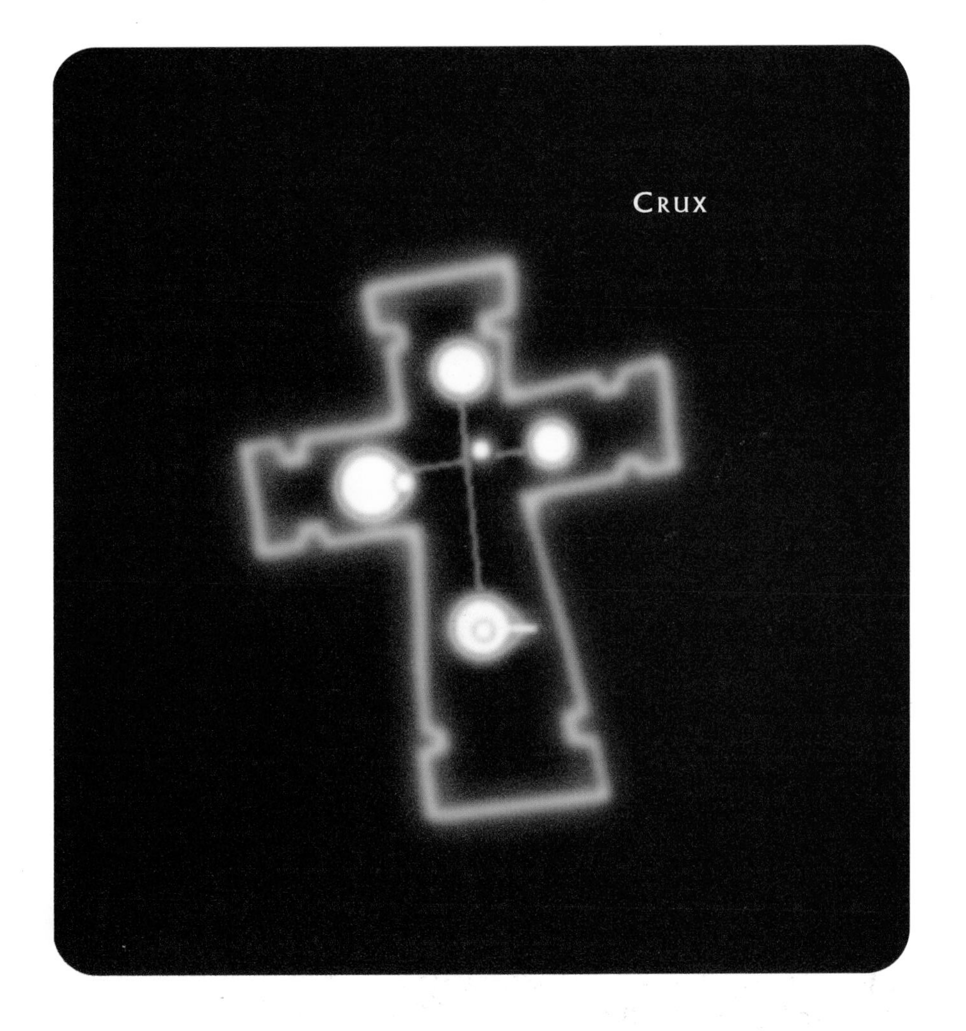

CRUX: The (Southern) Cross

CRUX is the smallest constellation in the night sky. It is a beautiful cross or kite-shaped pattern of stars, nestled in the southern Milky Way, and best seen during southern autumns. Along with the bright stars Alpha and Beta Centauri, the stars of Crux are a magnificent sight.

Even though it's small, the Southern Cross is one of the best known constellations. You see Crux on the flags of many countries in the Southern Hemisphere, including New Zealand, Australia, Samoa, and Micronesia.

The stars of Crux shine against the brilliant backdrop of stars of the Milky Way. But, if you look closely, you'll also see a dark part of the Milky Way touching the cross. This dark patch is called the Coalsack. It's a huge nebula of gas and dust. Unlike many nebulas that shine, this nebula blocks the light of more distant stars. That's why we see it as a black silhouette against the Milky Way.

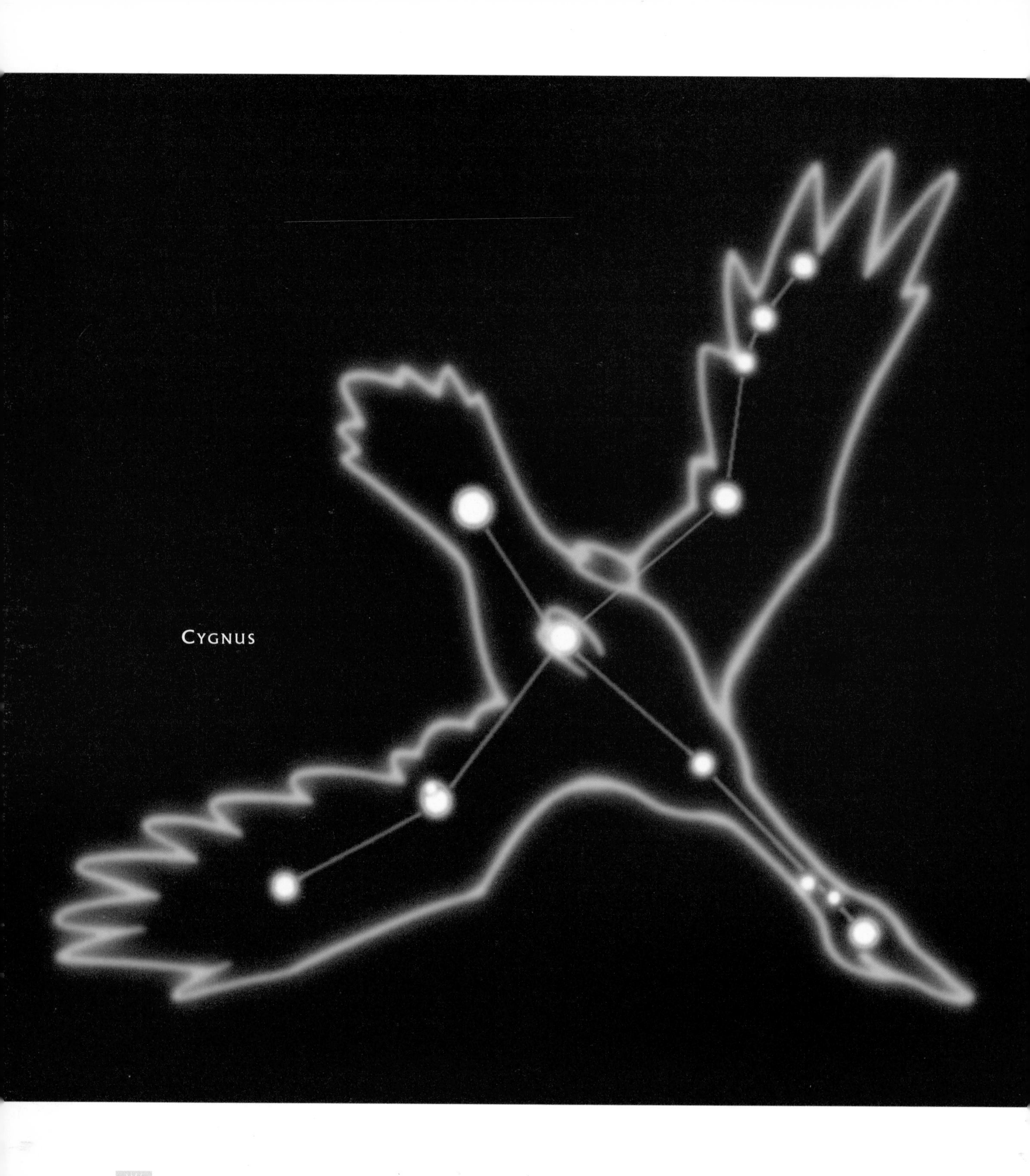
CYGNUS

CYGNUS: The Swan

CYGNUS the Swan is one of the best pictures in the night sky. It doesn't take a lot of imagination to see a great swan in this constellation, soaring along the Milky Way. The six brightest stars of Cygnus also make a lovely cross or kite shape.

The star in the tail of the swan is called Deneb. It shouldn't surprise you to know that *deneb* is the Arabic word for "tail." Deneb, along with Vega and Altair, form the Summer Triangle that stargazers in the north see all summer long.

The Coeur d'Alene Salishans, people who lived where the Great Plains meet the Rocky Mountains, see a goose in the stars of Cygnus. They tell a story of how the goose got into the sky.

One day, three hunters were hiking through the forest when they came upon a clear, deep lake. The lake was the home of many birds, including a beautiful snow goose. As the hunters emerged from the forest, they spotted the goose, drifting on the calm waters.

One hunter raised his bow but the others protested. They didn't want to kill such a beautiful bird. The goose was startled by the noise of the hunters and took flight. Despite his companions' cries, the hunter let fly an arrow. The arrow found its mark and the snow goose fell into the lake. It sank, deep into the dark, cold waters.

The sky darkened and, as the hunters searched the waters of the lake for the goose, they saw the stars reflected in the surface. In those stars, they saw a new constellation. It was the spirit of the snow goose, flying through the heavens.

Astronomers have discovered a very strange object in Cygnus. This object is about 8,000 light-years away, and it orbits a huge star every six days. That's fast. After all, the Earth orbits our Sun once every 365 days! The object is 10 times as massive as our Sun. Astronomers can't see it, but it gives off very strong x-ray radiation. They call it Cygnus X-1. What is it? It's a very unusual star called a black hole—a star that's so massive and with such strong gravity that nothing can escape from it, not even light. It's a star that no one can see.

DELPHINUS: The Dolphin

DELPHINUS may be one of the smaller constellations, but it's also one of the prettiest star patterns in the sky. There aren't many stars in this constellation, but they still manage to make a picture that looks a lot like a dolphin.

Delphinus swims near the Milky Way, not far from Pegasus, in the late summer sky.

Why is there a dolphin in the night sky? According to one myth, the sea god Poseidon sent the dolphin Delphinus on a mission—to find Amphitrite, a sea nymph, to be his bride. The dolphin completed his mission successfully and was rewarded with a place in the heavens.

Another story about the dolphin would make a great music video: 1,300 years ago, there was a Greek musician named Arion. Like a rock star today, he traveled to different cities to play his music. After a successful tour of Sicily, Arion was returning to Greece when his ship was attacked by pirates. They captured the ship and were ready to rob and kill Arion, when he made a last request. Of course, Arion's wish was to sing a final song.

The song that Arion sang was so lovely that it drew an audience of dolphins from the sea. Seeing his chance for escape, Arion jumped overboard. One of the dolphins, obviously a big fan of his music, carried Arion to safety and back home to Greece. This good deed didn't go unnoticed. Apollo, the god of music and poetry, rewarded the dolphin by placing him—and Arion's lyre—in the sky.

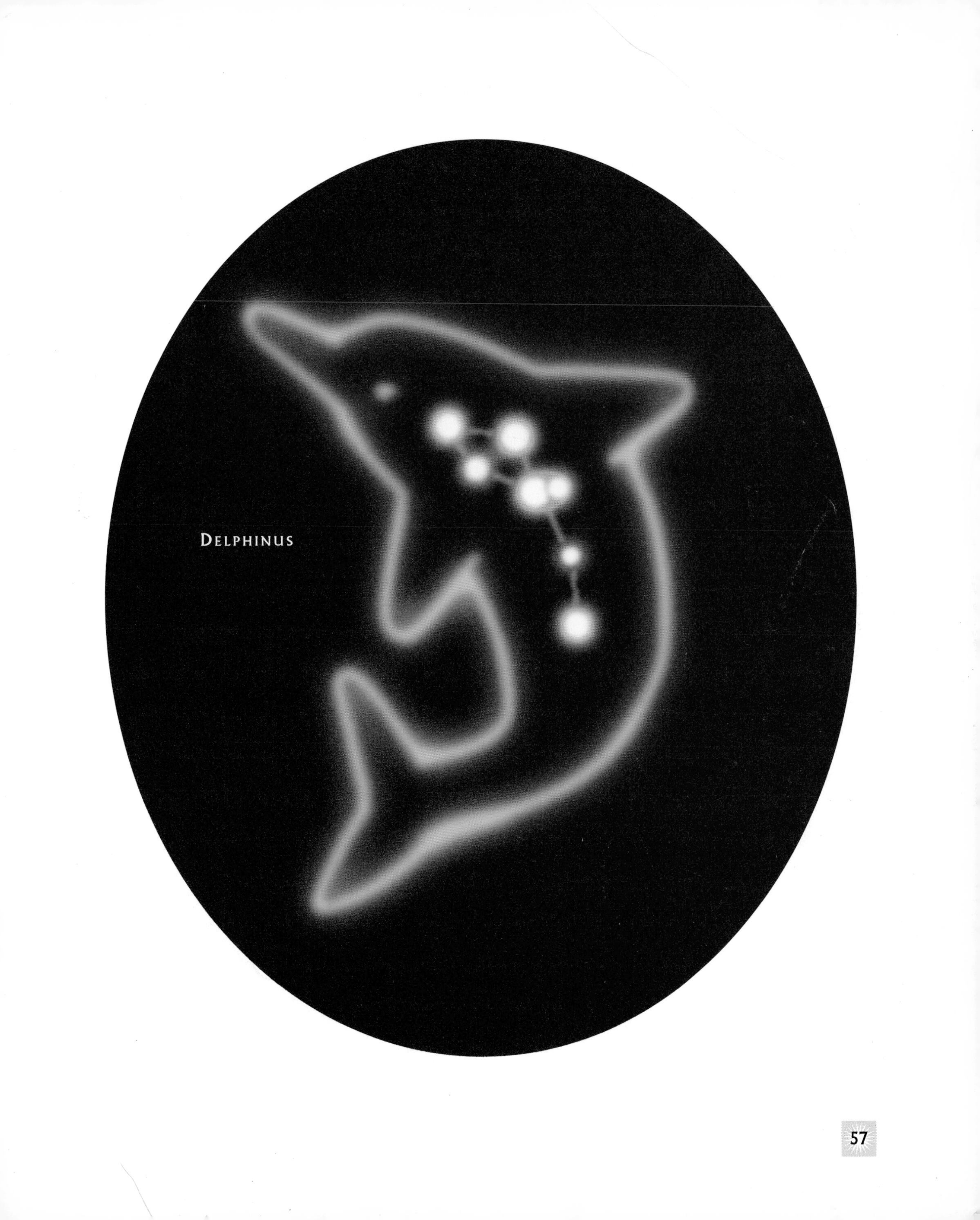
DELPHINUS

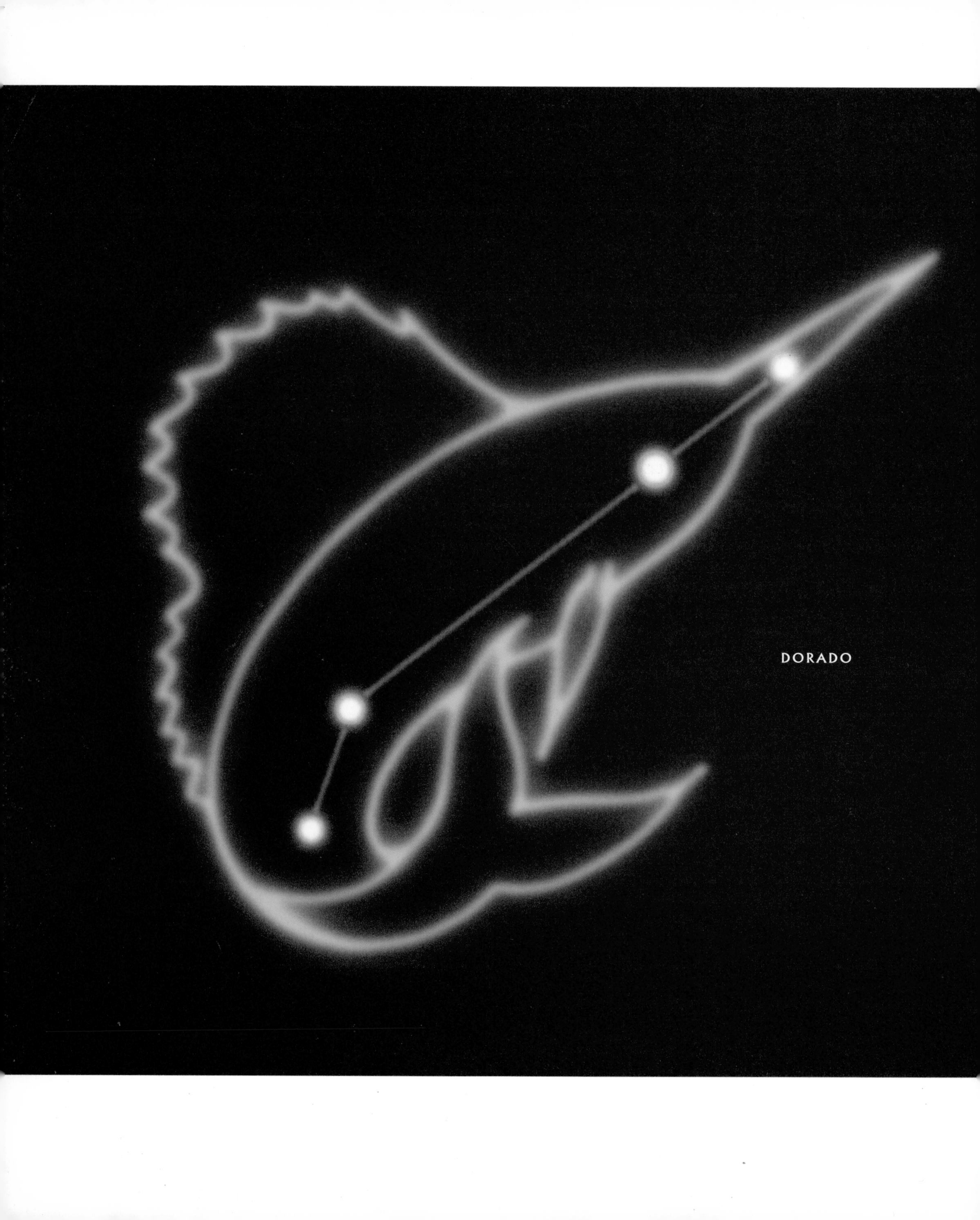
DORADO

DORADO: The Swordfish

DORADO is a very small constellation that southern stargazers can see most nights. But it's best seen during the southern summer.

It is one of the constellations invented by the Dutch navigators Pieter Dirkszoon Keyser and Frederick de Houtman over 400 years ago. Their constellations were of strange creatures, and included this tropical dolphinfish or swordfish.

Dorado may be small but it contains a very interesting sight. When you look up at Dorado, you'll see a patch of light that looks like a piece of the Milky Way. That patch of light is the Large Magellanic Cloud, an oddly shaped galaxy about 20,000 light-years wide and 170,000 light-years away. It's the nearest galaxy to our own. It orbits our galaxy the way the Moon orbits the Earth and the planets orbit the Sun.

There are no bright stars in Dorado, but in 1987, a brilliant star suddenly appeared in the Large Magellanic Cloud. It was a supernova, an exploding star. Before the explosion, you couldn't see the star without a telescope. After it became a supernova, it was so bright that you could see it without binoculars or a telescope for nearly a year until it grew dimmer. It really was super!

DRACO: The Dragon

DRACO the Dragon winds menacingly around the north celestial pole and Ursa Minor, the Little Bear. It is the eighth largest constellation and, even though it may not be made of bright stars, you can still picture the long line of stars as a dragon. Because it's so close to Polaris, you can see this creature every night of the year.

People have seen many different dragons in these stars. The Sumerians saw the serpent named Tiamat. Others saw the dragon that was slain by St. George.

According to Greek mythology, Draco guarded the Golden Apples of the Hesperides. You'd think that a dragon would be plenty of protection for some apples. But not against the warrior Hercules. The great hero killed the beast and picked the apples, thereby completing one of his Twelve Labors. That's why, on star maps, you sometimes see the triumphant hero with one foot on the dragon's head.

Today, Polaris is the star almost directly over the Earth's North Pole. But 5,000 years ago, a different star was the "north" star. That star was the brightest star in Draco—Alpha Draconis, or Thuban.

EQUULEUS: The Little Horse, The Colt, The Foal

NOT far from the great winged horse, Pegasus, is the small horse Equuleus. This is a really small horse. Equuleus is the second smallest constellation in the sky.

You can look for this pony in autumn skies, but there's not much to see.

Draco
Equuleus

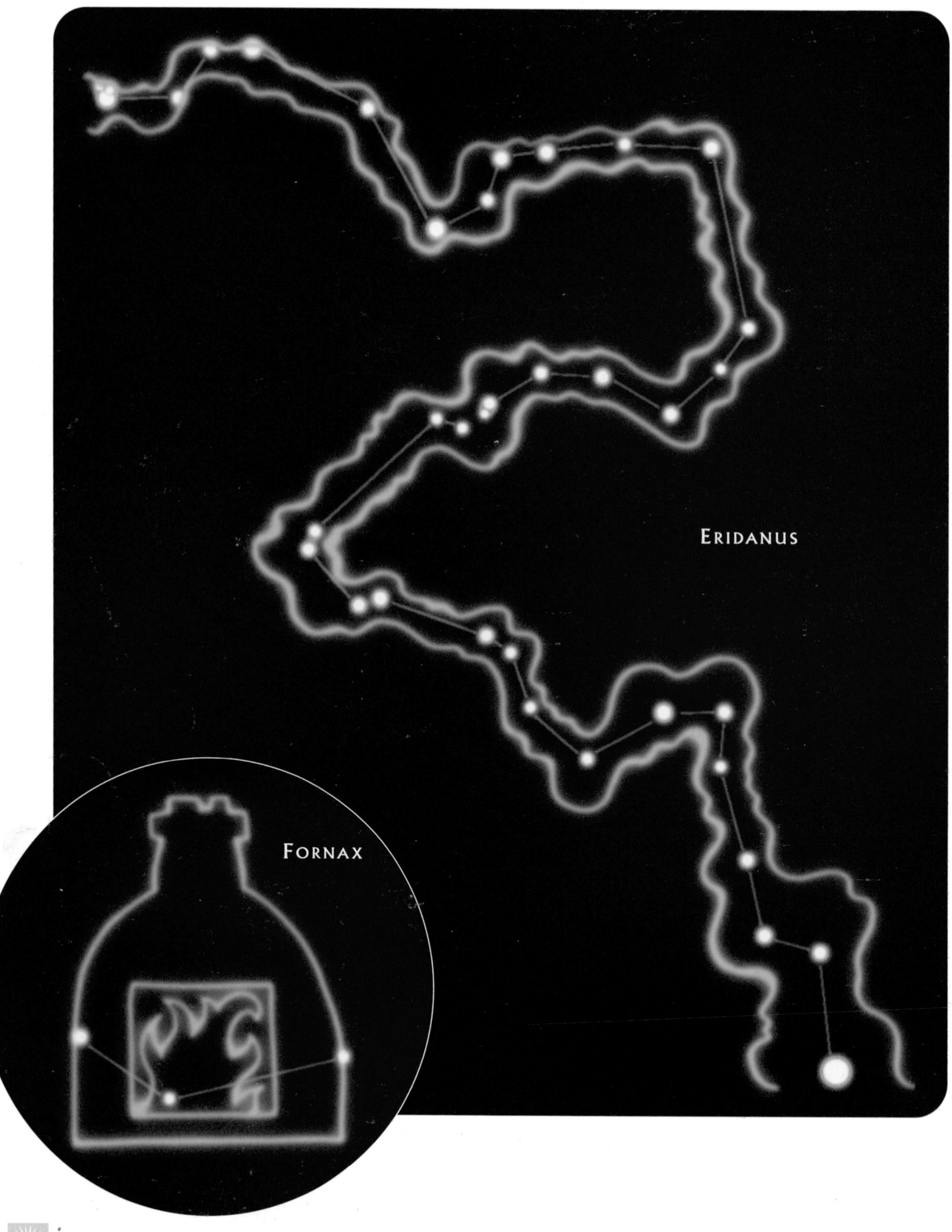
ERIDANUS
FORNAX

ERIDANUS: The River, The Celestial River

Eridanus flows through the night sky from November to January. It's the sixth largest constellation—a long, flowing string of stars that winds its way from the feet of Orion in the north, to Tucana the Toucan in the south.

The star at the south end of the river is Alpha Eridani. It's also known as Achernar, which is Arabic for "river's end."

People have called this constellation a river since ancient times. Some people know it as the Euphrates. To others, it's the Nile. And to some, it's the Po River in Italy.

In Greek mythology, the river plays a part in a very sad story.

Phaethon was the son of Helios, the Sun god. One day, Phaethon asked Helios if he could have the keys to his father's chariot and take it for a spin around the sky. Helios agreed, but not before he gave his son a long lecture on how to drive carefully.

Phaethon headed off in the chariot, but he soon found himself at a dizzying height. He was so high up, he encountered the creatures of the zodiac, who frightened him. He steered the chariot toward the Earth, but he then came too close to the planet. He swerved back up into the heavens but, this time, the stars complained about the young charioteer's reckless driving.

It was Zeus who finally put an end to Phaethon's joy ride, by causing him to crash into the River Eridanus.

One of the stars in this constellation is called Epsilon Eridani. Does the name ring a bell? This star is a lot like our Sun. It lies about 11 light-years away and is the tenth nearest star to us. Give up? According to *Star Trek*, Spock's home planet Vulcan orbits Epsilon Eridani!

FORNAX: The (Laboratory) Furnace

Fornax is a small constellation, with no bright stars, sitting on the shores of Eridanus, the River. As long as you live closer to the equator than the poles, you can see it high in the sky in November and December.

It is one of de Lacaille's constellations and is supposed to be a chemical or laboratory furnace. Like all of de Lacaille's constellations, Fornax has only been around for about 250 years.

GEMINI: The Twins

WITH just a bit of imagination, you can see the twins Castor and Pollux in the stars of Gemini. They stand beside each other in the winter sky, with their feet in the Milky Way. Gemini is a constellation of the zodiac, with Cancer on one side and Taurus on the other.

You'll find Gemini in the winter sky, with the twin stars marking one corner of the Winter Hexagon (see page 24).

In Greek mythology, Castor and Pollux were known as excellent sailors. In fact, they were Argonauts, members of the famous crew that sailed with Jason on his voyage to find the Golden Fleece. They did many things on the voyage that earned them their reputation, which explains why they're the patron saints of sailors.

On the voyage, Pollux even fought Amycus—one of the sea god Poseidon's sons. Amycus was a giant and, according to some, the world's greatest bully. When Jason and the Argonauts landed in his kingdom, the giant would not let them leave until someone agreed to fight him in a boxing match. Of all the Argonauts, it was Pollux who stepped into the ring with Amycus.

Amycus was a formidable foe. Not only had he invented boxing, but he usually killed his opponents with a single blow. The Jabbing Gemini, however, was more than equal to the task, being more agile and faster than his opponent. He defeated Amycus, and Jason and the Argonauts were free to continue on their way.

You can still find Castor and Pollux at sea today. During storms, sailors sometimes see an electrical glow in the rigging of their ship. It's called St. Elmo's fire. When there are two glowing lights in the rigging, the twin lights are called Castor and Pollux. Because the Gemini twins are the patron saints of sailors, the lights are seen as a sign of a safe and successful voyage.

Did you know that there are twins in the Twins? When you look at the bright star Castor through a small telescope, you'll see that it is actually two stars. These stars orbit each other. When they studied these two stars, astronomers discovered that each of them was really a pair of stars! That makes four stars altogether.

And that's not all. There is another, dimmer star nearby that is also part of this star system. You guessed it—this star is also a pair of stars orbiting each other. So, when you look at Castor, you are actually looking at three sets of twin stars in the Gemini Twins!

Geminid meteors shower down from Gemini around December 14th. Imagine, fireworks in the winter!

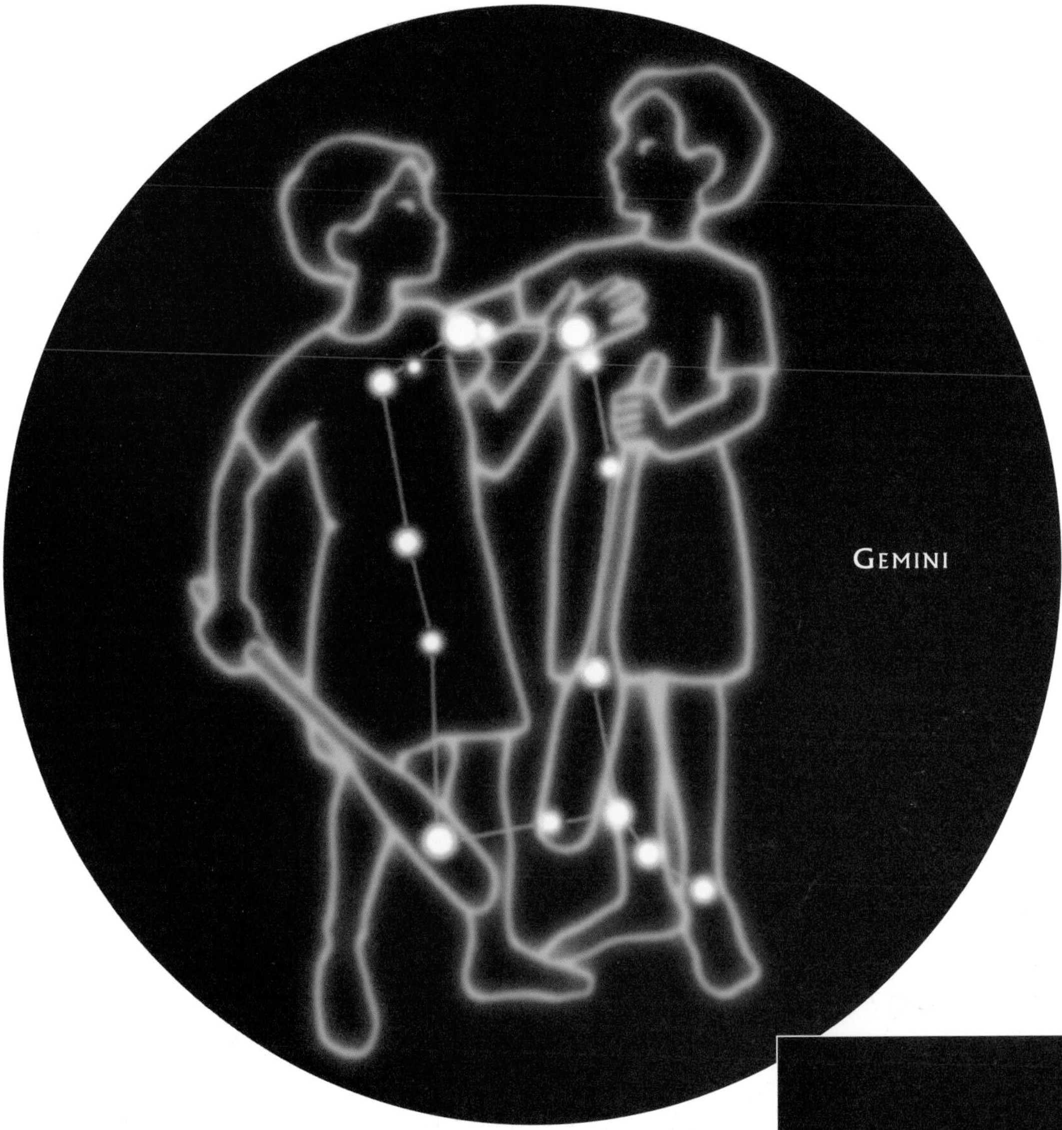

GEMINI

GRUS: The Crane

GRUS is one of the many birds flying among the stars. Like Cygnus the Swan, Grus is made up of a cross-shaped pattern of stars. This bird soars through southern skies, and is flying high from September to October.

The constellation was invented by Keyser and de Houtman over 400 years ago. The crane is one of the many exotic creatures that these two Dutch navigators placed in the sky.

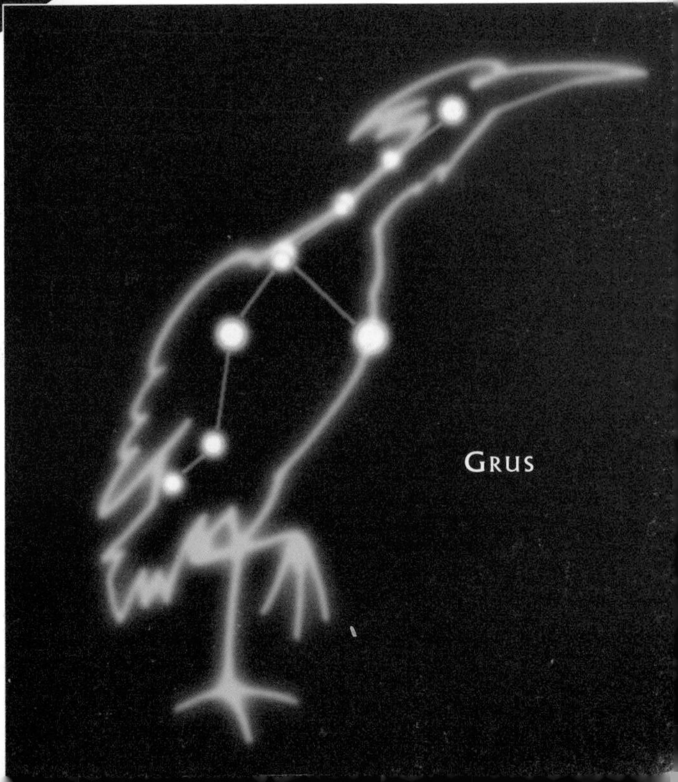

GRUS

HERCULES: The Strongman, The Hero

THE fifth largest constellation sprawls across the sky on summer nights, standing off to one side of the Milky Way between the very bright stars Vega and Arcturus. It is Hercules, a constellation that looks a bit like a crooked H with arms and legs. Four of the stars in the crooked H—two in Hercules' lower torso and two in his upper thighs—form an asterism named the Keystone.

Hercules, who is also known as Heracles, is one of greatest heroes of all time. He had many adventures, but he's famous for completing the Twelve Labors—some of which are portrayed every night in different constellations. Hercules' labors included slaying a lion, a many-headed serpent, and an entire flock of birds that were ravaging the countryside.

He had to kill a dragon to pick the Golden Apples of the Hesperides. He had to capture a wild boar, a horned animal called a hind, four horses, a bull, an entire herd of cattle, and a devilish dog. He even had to clean a king's stables, and find a queen's girdle. No wonder he gets his own constellation!

On a clear, dark night, look closely at Hercules' right hip and you'll see something that looks like a fuzzy star in the Keystone. This object is called M13. It's really a collection of stars called a globular cluster—an enormous, sphere-shaped grouping of hundreds of thousands of stars.

There are many globular clusters in the sky, but M13 is special. In 1974, scientists used a powerful transmitter to send a radio signal toward M13. The signal was meant for any alien civilization that might live on a planet orbiting one of the stars in M13. The message included information about our solar system, our planet, and us.

Have any aliens received the signal? Have they answered with a signal of their own? We'll have to wait a long time before we know the answers to these questions. M13 is 25,000 light-years from Earth. That means that our radio signal, traveling at the speed of light, won't reach M13 for another 25,000 years. And, even if someone is there to answer our signal, their reply would take another 25,000 years to reach us. So, will our radio signal be answered? We'll find out in 50,000 years. And you thought the mail was slow!

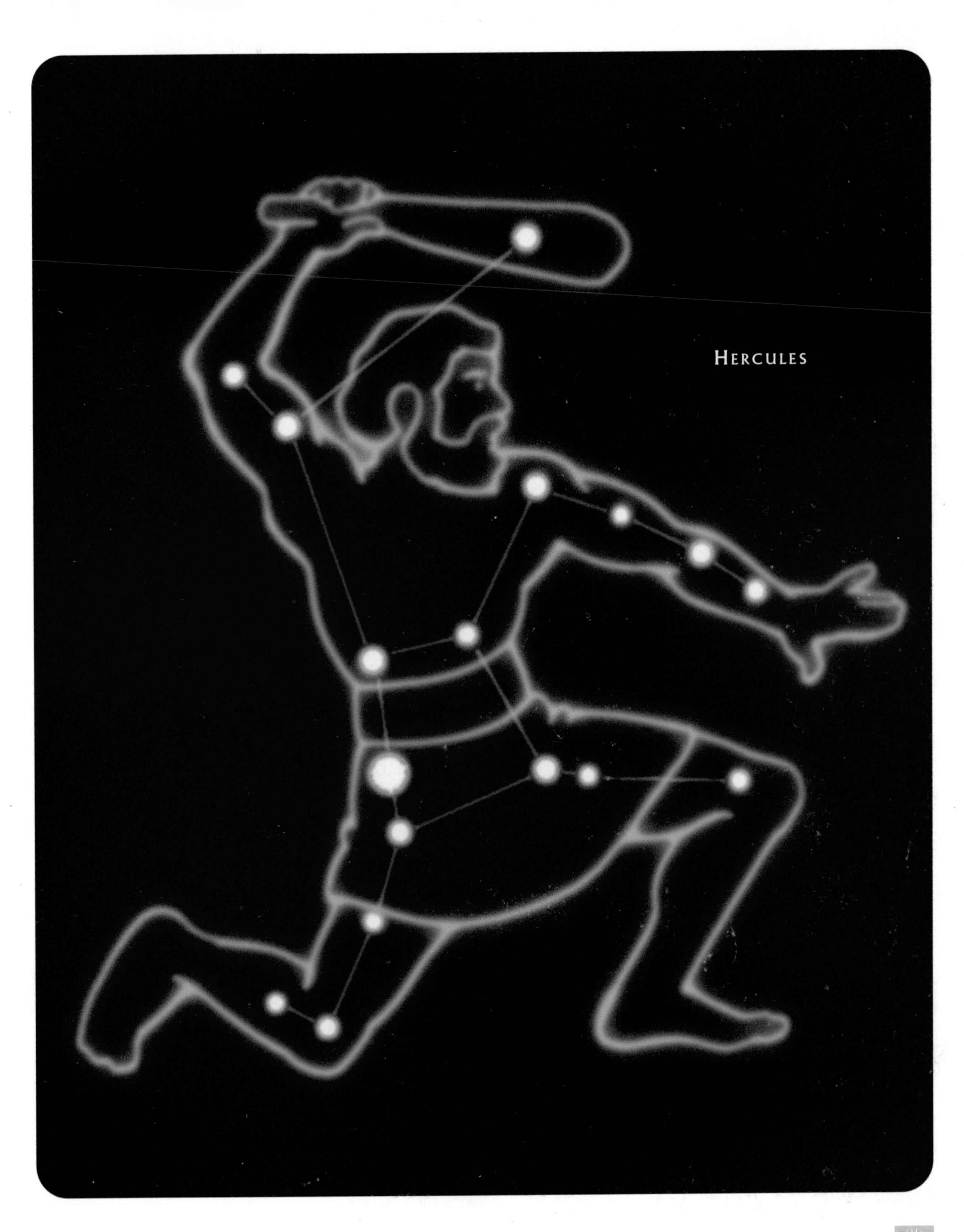
HERCULES

HOROLOGIUM: The Pendulum Clock

HOROLOGIUM is a fairly small constellation, with no bright stars. You can see it in southern skies during the southern spring and early summer.

One of the constellations invented by de Lacaille, it looks like a pendulum clock—a clock that keeps time with the steady back-and-forth motion of a pendulum.

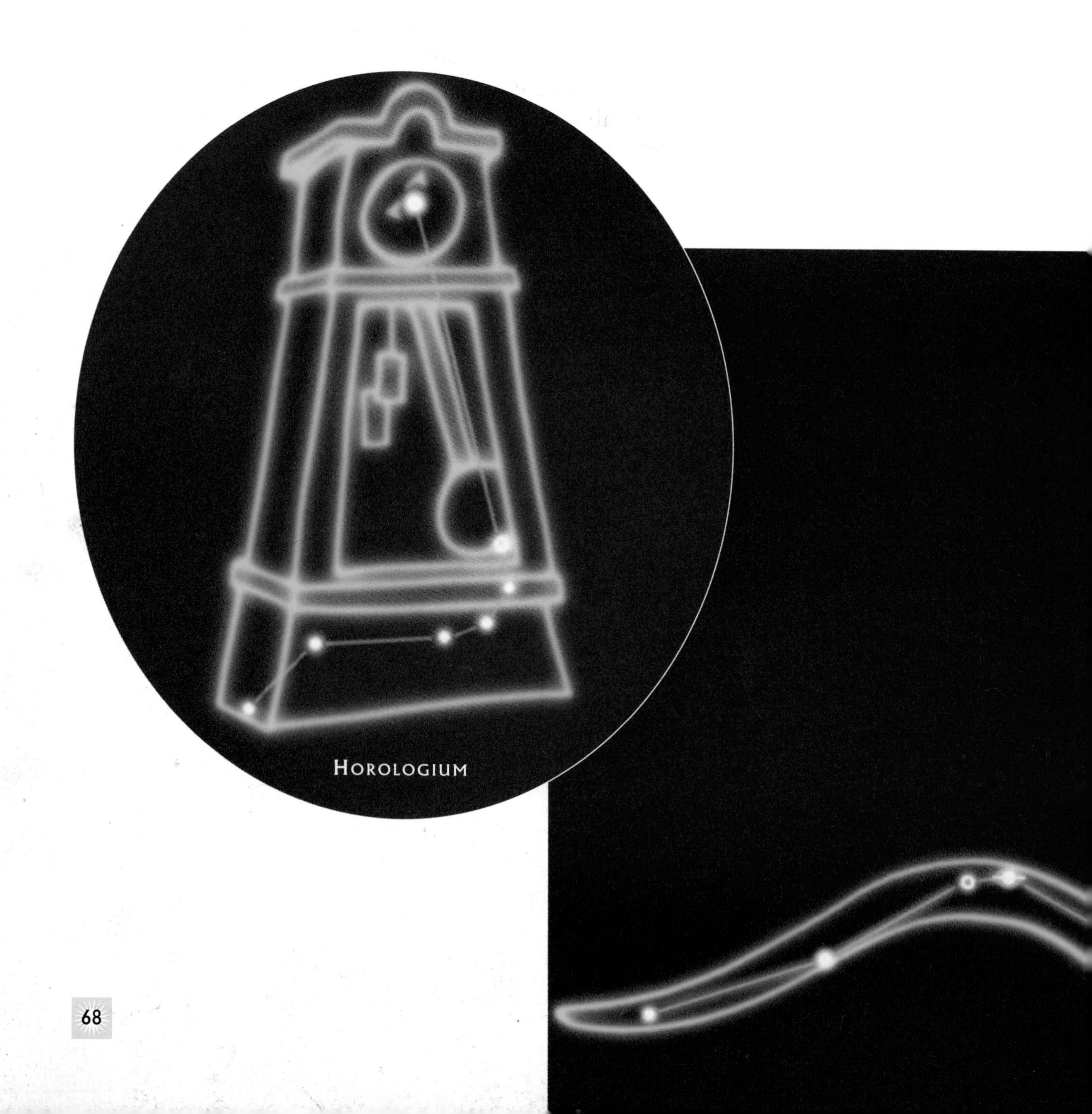

HYDRA: The Sea Serpent, The Water Snake

HYDRA is the largest and longest constellation, a serpent of stars that stretches almost a quarter of the way around the sky, from Cancer to Centaurus to Libra. It is so long, it can take more than six hours to rise completely into the sky. You can see Hydra during the southern autumn months.

To the Greeks, Hydra was a monster, a serpent-like creature with many heads. Some stories say Hydra had five or six heads, others say it had a hundred. Fighting the creature was the second of Hercules' Twelve Labors. And what a challenge it was! Every time Hercules cut off one of Hydra's heads, another would grow back in its place. But even when Cancer the Crab joined the battle against the great hero, Hercules was victorious, killing Hydra with arrows dipped in the creature's own venom.

HYDRUS: The Southern or Little Water Snake

HYDRUS is a small constellation, not far from the south celestial pole. It is a line of stars that is visible on most nights, zigzagging between the Large and Small Magellanic Clouds.

It belongs to the zoo of creatures placed in the sky over 400 years ago by the Dutch navigators Keyser and de Houtman.

INDUS: The American Indian

DON'T expect to see a human figure when you look at the stars of Indus. There aren't any bright stars and the pattern they make doesn't look anything like what you'd expect. Indus isn't far from Sagittarius and is best seen in southern skies from August to October.

This is another of the constellations created by Keyser and de Houtman over 400 years ago. Like the Greek myths and other stories in the stars, this constellation shows us how people thought hundreds of years ago. Keyser and de Houtman might have put Indus in the sky because people back then knew very little about native North Americans. The two astronomers might have thought these people were exotic and strange, and belonged in the sky with creatures like unicorns and dragons. Today, of course, if anyone invented a constellation, they wouldn't call it the American Indian.

LACERTA: The Lizard

LACERTA, like the creature it depicts, is a small constellation that is easy to overlook. It sits in autumn skies, between Andromeda and Cygnus, and slithers across the Milky Way.

Can you imagine a tiny lizard in this collection of stars? That's what the German astronomer Johannes Hevelius saw when he invented this constellation in the late 17th century.

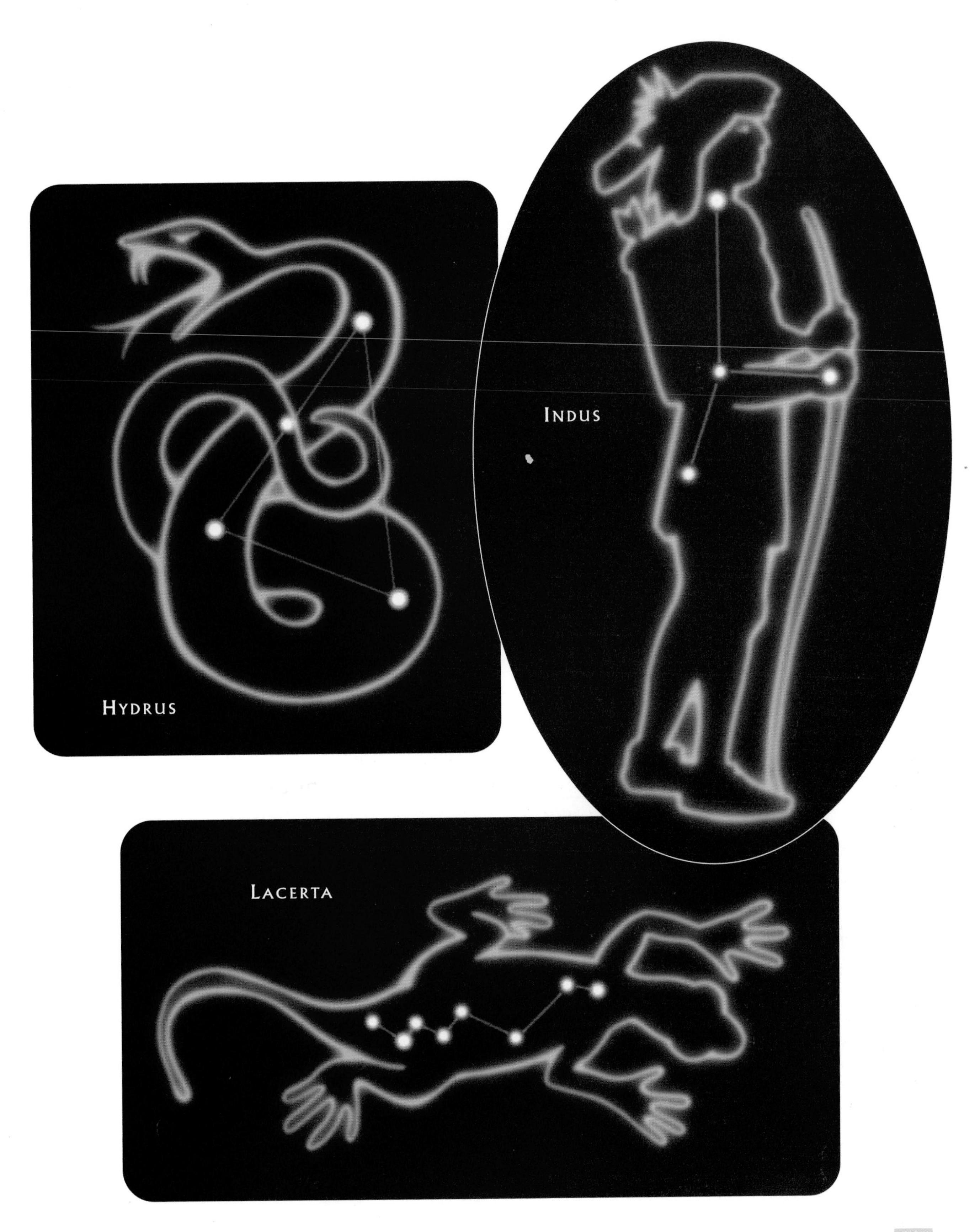
Hydrus
Indus
Lacerta

LEO: The Lion

LEO is a constellation of the zodiac and one of the better star pictures in the sky. It isn't hard to see a lion's head, mane, and body in this pattern of stars. An asterism, called the Sickle, looks like a backwards question mark and is the lion's mane. A bright star marks the lion's heart. It's called Regulus, which means "the Little King." At the other end of the lion is the bright star Denebola, which means—what else?—"lion's tail."

The best time to see Leo is during the spring. If you can't find the lion right away, find the Big Dipper first. Then imagine poking a hole in the bowl of the dipper. The water that leaks from the bowl will drip right onto the lion.

Not surprisingly, a lion has been seen in these stars for thousands of years—by such people as the Sumerians, Babylonians, Persians, Syrians, and Romans. But to the Chinese, the stars of Leo were a horse instead. And, to the Inca sky watchers, they were a puma.

To the Greeks, Leo was the Nemean Lion, a monstrous creature that tormented people and ate their herds. The lion's skin couldn't be pierced by arrows or swords, so it took a hero like Hercules to kill it. At first, Hercules tried killing the Nemean Lion with his bow and arrow, but his weapons were useless. So, the great warrior battled the lion with his bare hands. By now, you probably won't be surprised to hear that Hercules was victorious. The lion was killed and placed in the sky so that everyone would remember Hercules' prowess.

Every year, around November 17, a shower of meteors radiates from this constellation. They're known as the Leonids. Sky watchers are very interested in watching this meteor shower, because every 33 years or so, there can be many, many Leonids. How many? If you had been watching this shower in 1966, you would've counted thousands of meteors in one hour!

LEO MINOR: The Little Lion, The Lion Cub

LEO isn't the only lion in the sky. There's also a lion cub, named Leo Minor. Unlike Leo Major, this constellation doesn't really look like a little lion. It's just a small pattern of stars that can best be seen in the spring, sitting between Ursa Major and Leo Major.

This constellation was one of seven invented by the Polish astronomer Johannes Hevelius over 300 years ago.

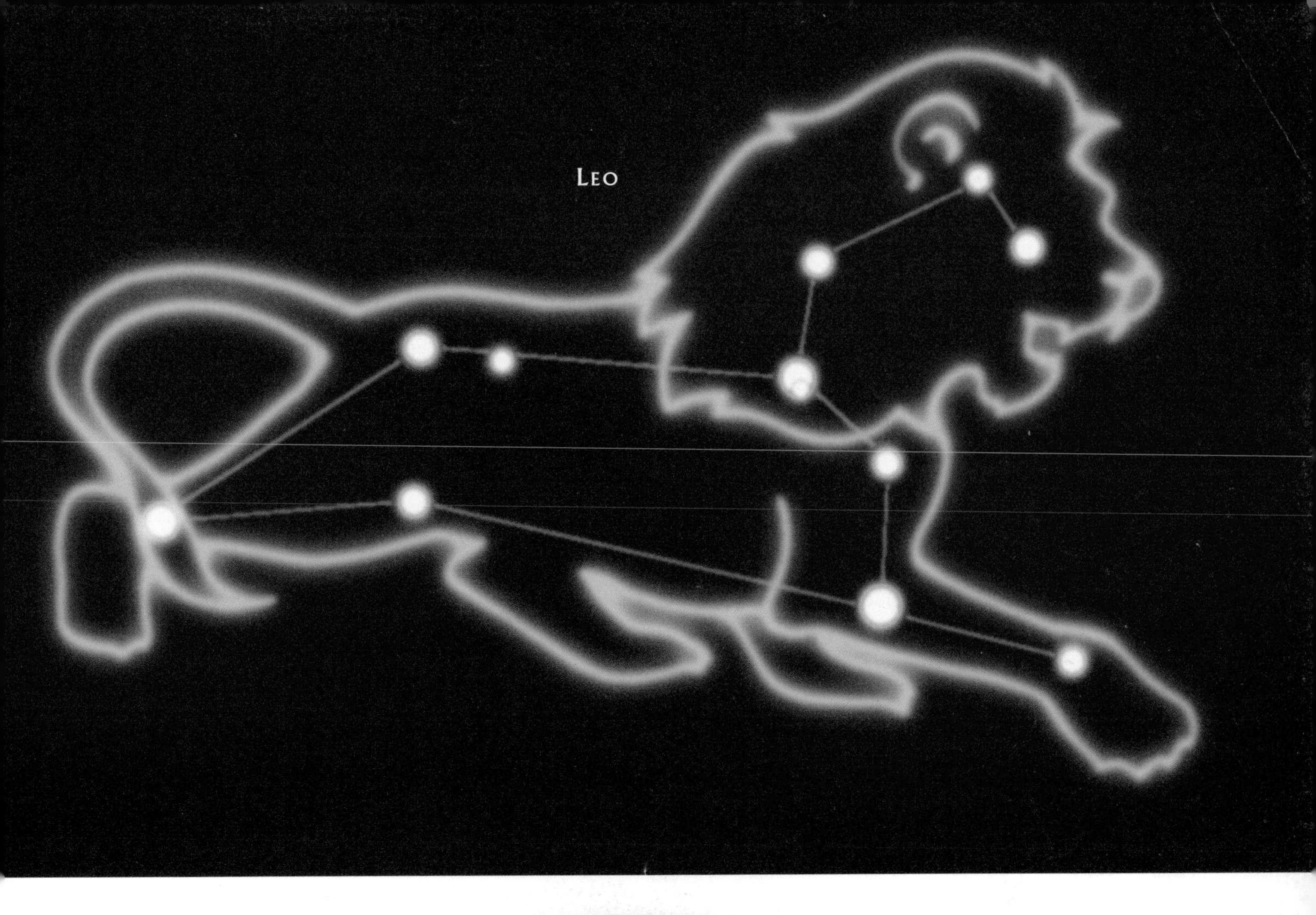
LEO

LEO MINOR

LEPUS: The Hare

LEPUS must be an arctic hare, because we see it on winter nights. It's a small constellation, with no bright stars, and can be as hard to find as a real hare hiding in tall grass.

They say that Lepus is a favorite prey of Orion the Hunter. If so, Orion doesn't have to look far for his quarry because Lepus crouches at the feet of the Hunter. On some star maps, you'll even see the hare being chased by Orion's dog, Canis Major.

Because the stars in Lepus don't look anything like a hare, it's not surprising that different people saw different things in these stars. Early Arab sky watchers saw four camels drinking from the River Eridanus. In Egypt, the constellation was a throne, or the ship in which the god Osiris sailed down the Nile River. To the Bororo people of Brazil—who had never seen a hare—the stars of Lepus, Orion, and Taurus were a giant alligator. Do you think Orion's dog would be so brave if he knew he were chasing an alligator instead of a hare?

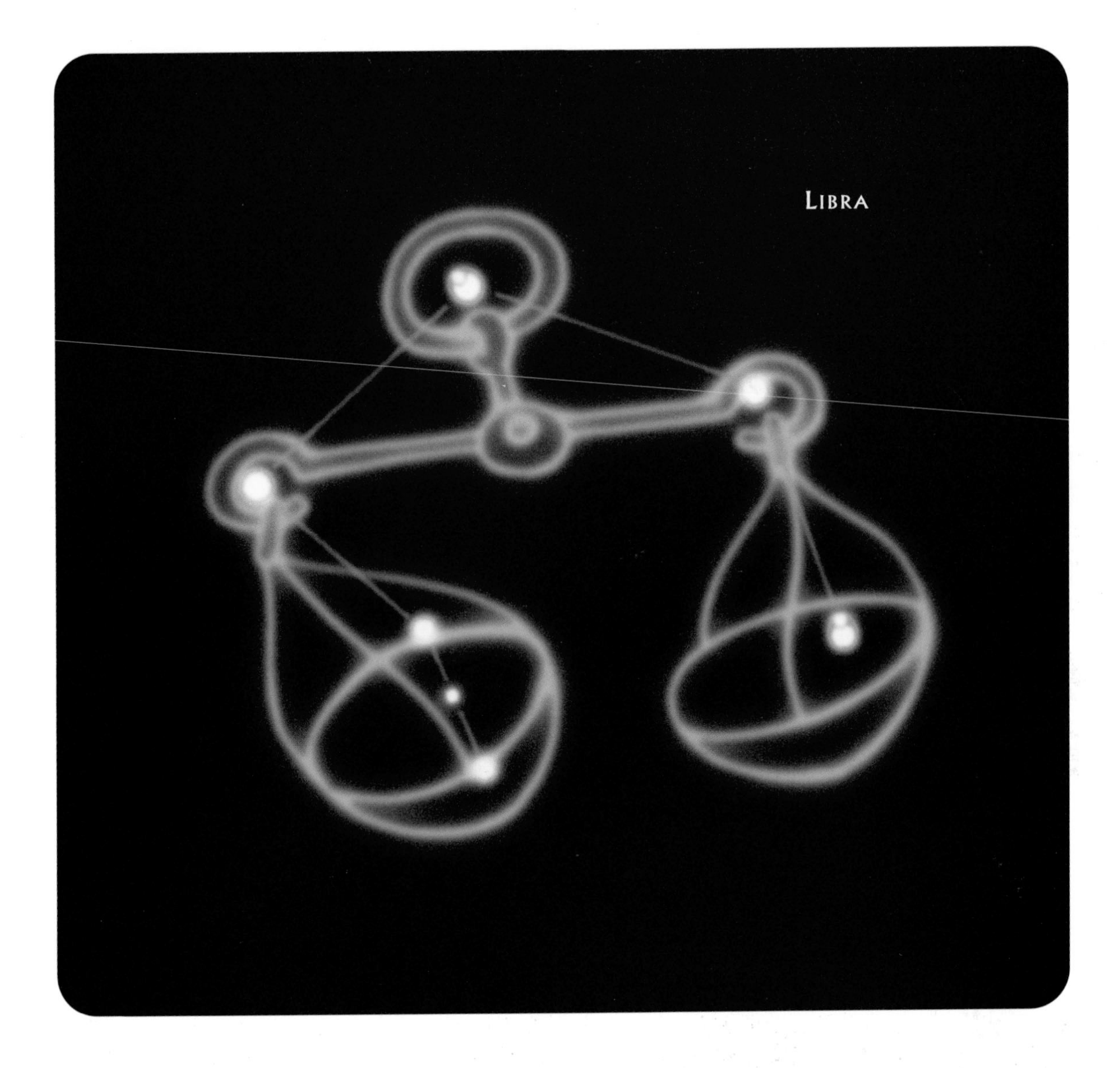

LIBRA: The Scales, The Balance

OF the 12 constellations of the zodiac, there is only one that is an inanimate object.

Libra is a set of scales, the old-fashioned kind that balances one side against the other in order to weigh things. It is seen in the late spring or summer skies, sitting between Virgo and Scorpius.

The two brightest stars in Libra are called Zubenelgenubi and Zubeneschamali. Zubenelgenubi means the "southern claw" and Zubeneschamali means the "northern claw." Why? Because, before it was known as Libra, this constellation was part of Scorpius the Scorpion—the part that pinches!

LUPUS: The Wolf

IT takes a lot of imagination to see a wolf in these stars. They're best seen in the late fall skies of the Southern Hemisphere, between Scorpius and Centaurus.

There's nothing much to see in Lupus today. But in the year 1006, you would've seen a bright supernova—an exploding star shining from the wolf.

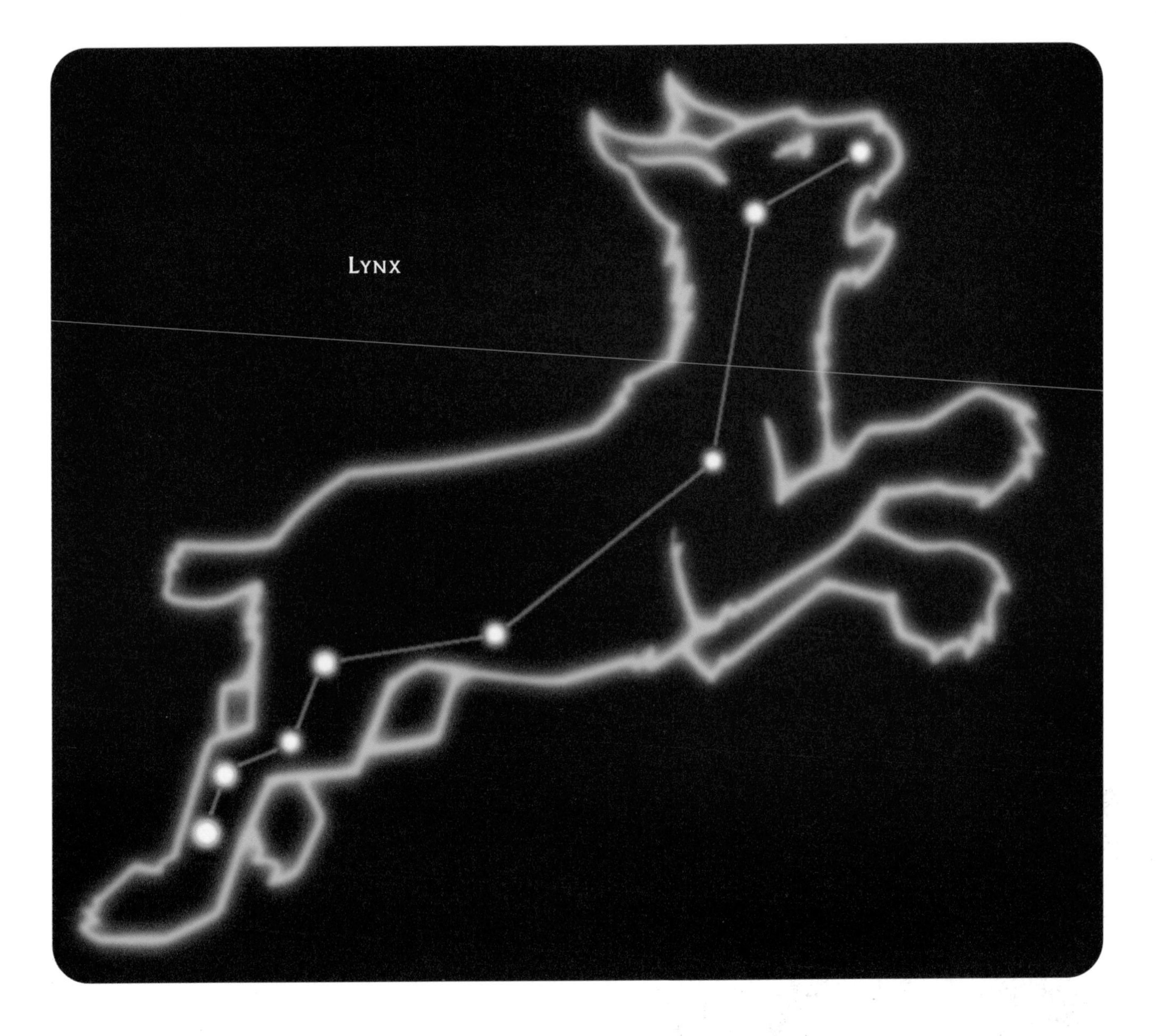

LYNX: The Lynx

EVEN though this constellation is bigger than Gemini, there are no bright stars and nothing much to see here. This cat prowls the winter skies between Ursa Major and the Twins.

It is one of the constellations created over 300 years ago by Hevelius. He called it the Lynx, because he said you would need the eyes of a lynx in order to see it.

The Lynx may be a big constellation, but it's losing stars! A star called 41 Lynx is moving through space very quickly. Three hundred years ago, it was in the constellation of the Lynx. Since then, the star has been quick as a cat and can now be found in the constellation next to it, Ursa Major.

LYRA: The Lyre, The Harp

YOU may be surprised to find only one musical instrument in the heavens—Lyra the Harp. It's not hard to see a harp in this constellation, made up of a triangle and a trapezoid.

Lyra can be seen high in the sky during the summer. Its brightest star, Vega, is one of the brightest stars in the sky, and marks one of the corners of the asterism called the Summer Triangle.

The Lyre in the sky is the mythical lyre that belonged to Orpheus, who was a great musician. His playing and singing tamed wild beasts. Trees and plants bowed down to him. Orpheus' music brought his wife, Eurydice, back from death, and even helped save the lives of Jason and the Argonauts.

One of the many dangers the Argonauts faced were the Sirens. A siren was a sea demon who was half woman and half bird. The sirens would tempt sailors with their magical music. The sailors, unable to resist, would steer toward the demons' island where the ship would run aground and all on board would perish.

The Argonauts would have sailed to the same fate, if it weren't for Orpheus. As the ship approached the demons' island and the song of the sirens became hard to resist, Orpheus began to sing. He sang so beautifully that the Argonauts were able to ignore the temptation and continue their voyage.

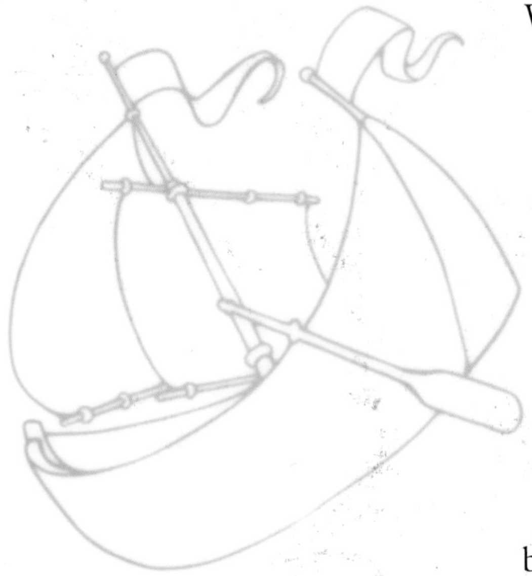

Orpheus was quite a star, and so is Vega. We say that Vega is a zero magnitude star. Magnitude is the way astronomers describe how bright a star is. Zero magnitude stars are brighter than first magnitude stars. First magnitude stars are brighter than second magnitude stars, and so on. We can see stars as dim as sixth magnitude with our naked eyes. The brightest star in the sky, Sirius, is so bright, astronomers say it has a magnitude of –1½.

About every April 22, you can see the Lyrid meteor shower, radiating from Lyra. Usually, you don't see a lot of Lyrid meteors, but it hasn't always been like that. In 687 B.C., Chinese stargazers, watching the Lyrid shower, described it by saying that "stars fell like rain." And, in 1982, people saw hundreds of meteors in only a few minutes. Wouldn't you be surprised one of these nights, to see meteors "fall like rain"?

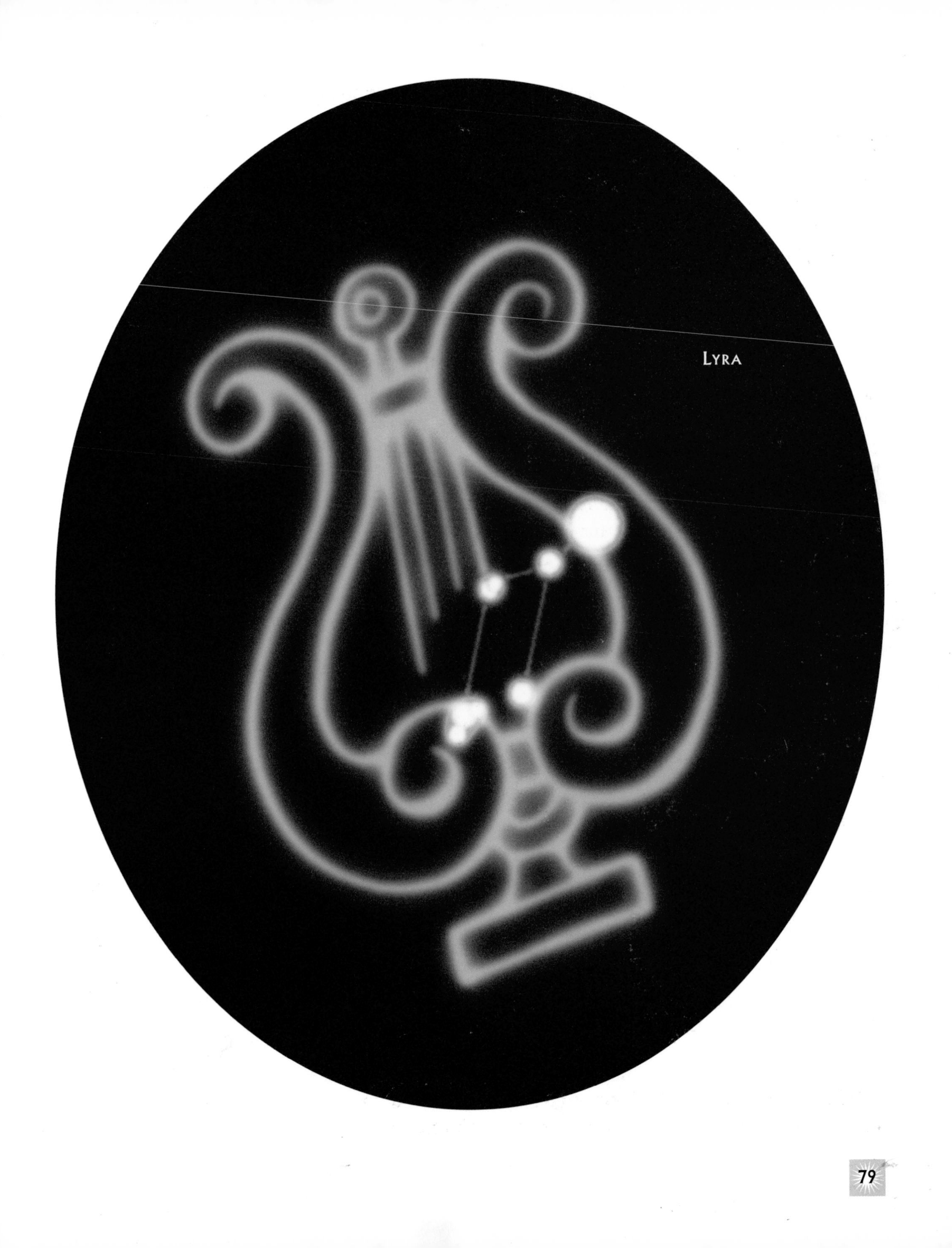
LYRA

MENSA: The Table Mountain

MENSA is a small constellation, near the south celestial pole. It doesn't have any bright stars, but it does contain part of the Large Magellanic Cloud, a small, oddly shaped galaxy.

Mensa was invented by the French astronomer Nicolas Louis de Lacaille. He named the constellation after a mountain near Cape Town, South Africa, where he had built his observatory over 250 years ago.

MICROSCOPIUM: The Microscope

THERE'S a telescope in the sky named Telescopium. It's for looking at things that are far away. There's also a microscope, named Microscopium, for looking at really small things. You almost need a microscope to see this constellation. It is small, with no bright stars. It sits near Telescopium and Sagittarius, in late summer skies.

Microscopium is another constellation invented by de Lacaille in the 1750s.

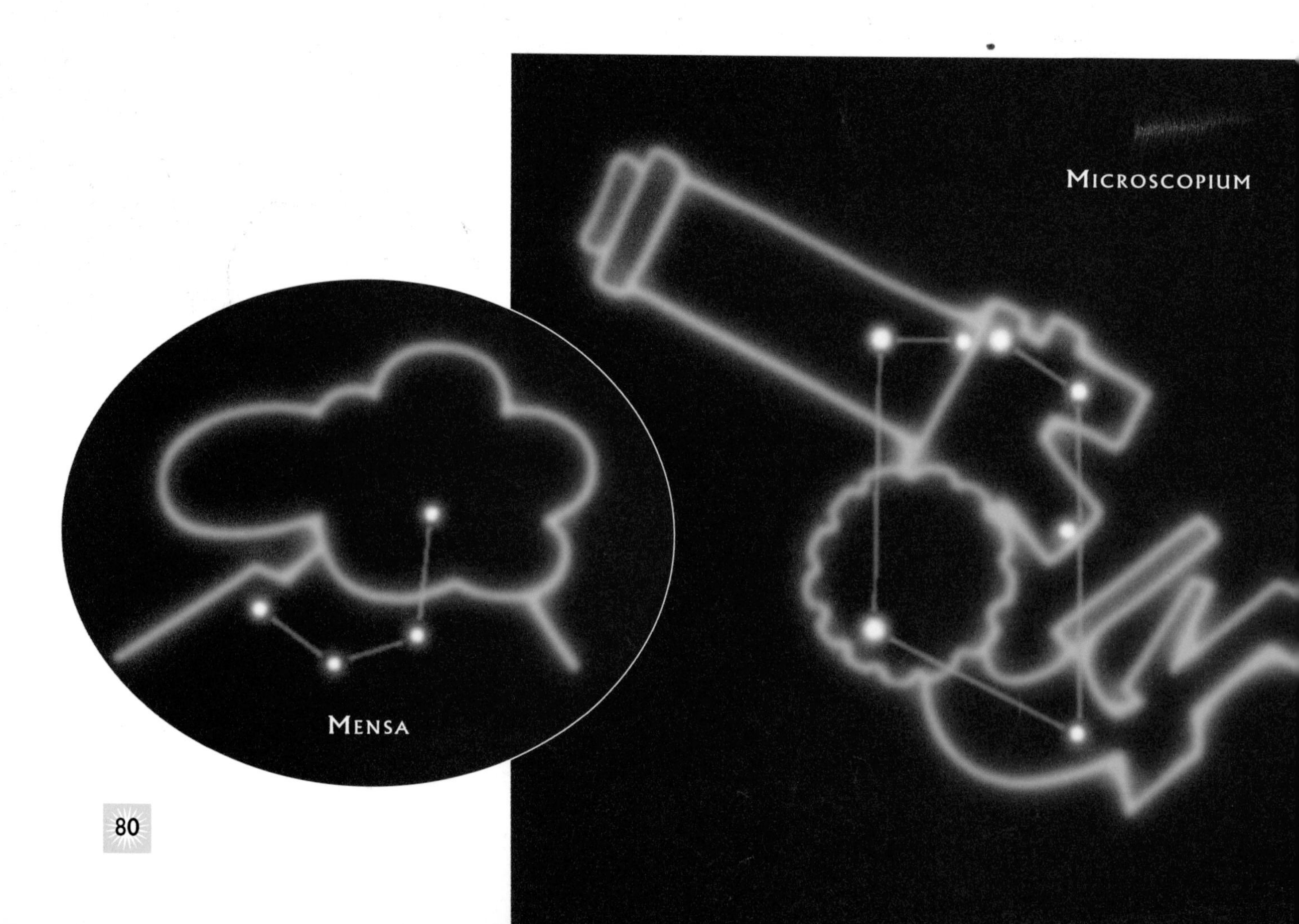

MONOCEROS: The Unicorn

MANY magical creatures are in the sky, but one of the most wonderful is Monoceros, the Unicorn. It gallops through winter skies, between Orion and Canis Major, and through the Milky Way.

This constellation was named nearly 400 years ago, by a Dutch mapmaker named Petrus Plancius.

One of the most amazing nebulas in the sky is in Monoceros. The Rosette Nebula is a huge cloud of gas, 5,500 light-years away. It looks like a beautiful, crimson rose, blooming in space.

MUSCA: The Fly

WHO would have thought you'd find a fly in the sky? But there it is, Musca the Fly, buzzing through southern skies. It's not far from Crux and Centaurus, and sits as if stuck in the web of the Milky Way.

It is one of the 11 constellations invented by Dutch navigators over 400 years ago.

NORMA: The Carpenter's Square

NORMA is a very small constellation in southern skies. It lies in the Milky Way, not far from Alpha Centauri.

Not surprisingly, there are no myths or stories to tell about Norma. It is one of the constellations that de Lacaille invented 250 years ago.

OCTANS: The Octant

OCTANS is an octant, an instrument used by navigators. It's a medium-sized costellation, but without any bright stars or interesting astronomical objects in it.

Like the other constellations named after instruments, it is one of de Lacaille's inventions.

If you stood on the Earth's South Pole and looked straight up, you would see Octans. The star closest to the south celestial pole (that is visible to the naked eye) is called Sigma Octantis. But it's not as close to the south celestial pole as the North Star is to the north celestial pole. It's not as bright as Polaris either.

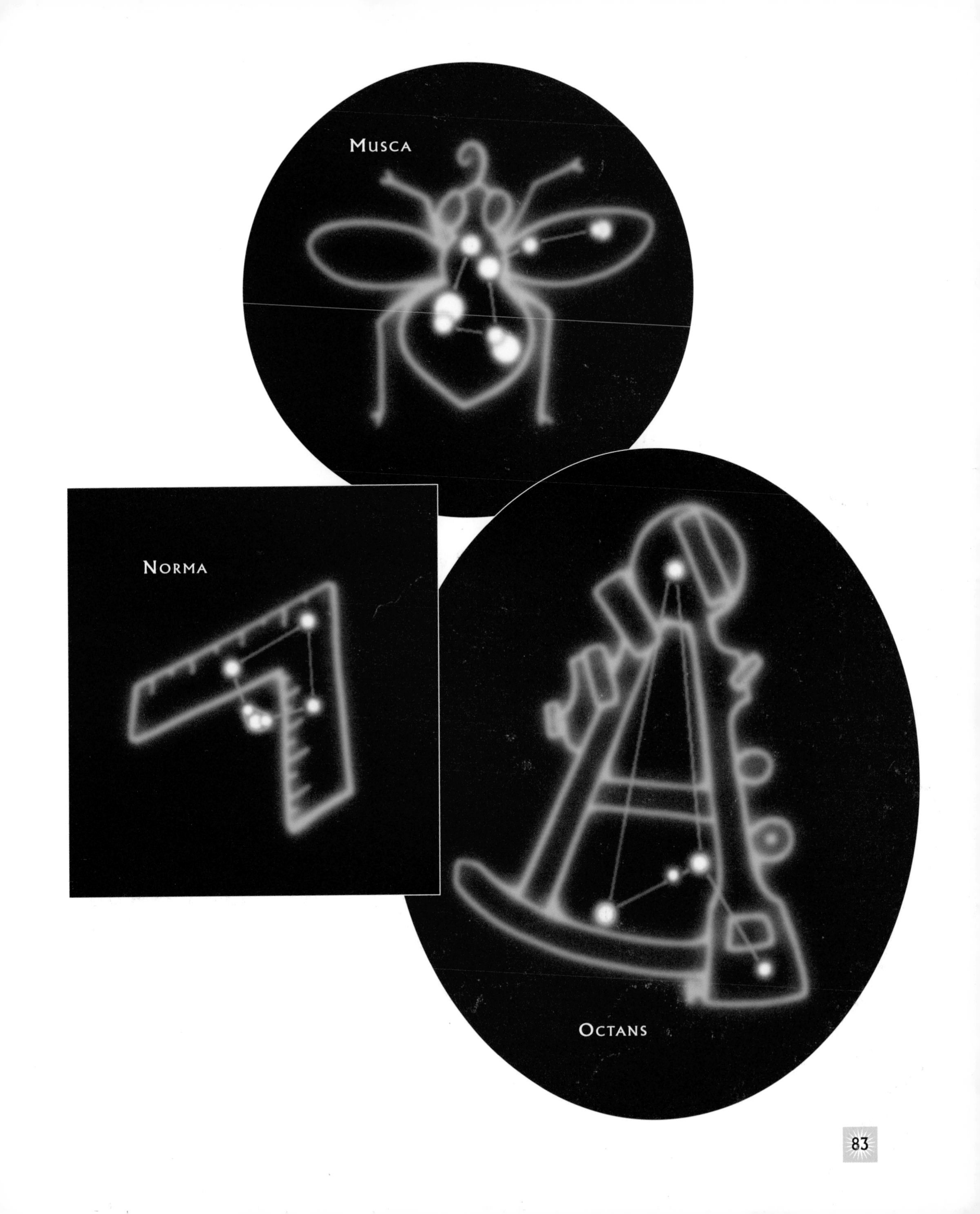
Musca
Norma
Octans

OPHIUCHUS: The Serpent Bearer

IT'S not surprising that a constellation of a god wrestling with a giant serpent is a large one. But Ophiuchus doesn't contain many very bright stars, so it isn't the easiest constellation to find. It's there though, in summer skies, standing between Hercules, Sagittarius, and Scorpius.

According to legend, this serpent bearer was also called Aesculapius. He was the god of medicine, and was such a good healer that he could bring the dead back to life. You'd think everyone would have liked Aesculapius because of this, but not everyone did. Pluto, god of the underworld, was not happy that Aesculapius was stealing subjects from him. So, at Pluto's request, the great god Jupiter took Aesculapius and placed him in the night sky, where we still see him.

You may have heard of one of the stars in this constellation. Barnard's Star is the nearest star to us after the Sun and Alpha, Beta, and Proxima Centauri. It's only six light-years away. It is also the star that is moving fastest through the night sky. Astronomers call this motion across the sky "proper motion," and Barnard's Star has the greatest proper motion of all the stars. It changes position by the width of the full Moon every 175 years.

Stars have proper motion because they're moving through space. Does this mean the constellations are going to look different every time we go stargazing? No. If you stand by the side of a highway and look at the cars going by, they zoom by very quickly. But if you look at cars moving on a highway very far from you, they look like they're moving slower—just because they're farther away.

Stars move through space, but most are SO far from us that we barely see them move. It would take hundreds or thousands of years before we saw most stars change position in the sky. Still, if you were to travel thousands of years into the future, the Big Dipper would look different. In fact, it wouldn't look like a dipper at all!

Ophiuchus

ORION: The Hunter

ALONG with the Big Dipper, Orion the Hunter is one of the best star pictures in the sky. It is a large constellation, standing in winter skies, with many bright stars. Some people think the stars of Orion form the shape of a crooked H or an hourglass.

On star maps, two bright stars mark the shoulders of the hunter and two mark his legs. One of the stars in his shoulders is called Betelgeuse. One of the stars in his legs is Rigel, part of the Winter Hexagon. The line of three stars is known as the Belt of Orion.

In Greek mythology, Orion was a boastful, vain warrior. He was so vain that he refused to believe that seven sisters known as the Pleiades could reject his romantic advances. And so we see him in the night sky, chasing them as he has for centuries. According to myth, Gaia, the goddess of Earth, became fed up with Orion's boasts that he could kill any of her beasts. She sent a deadly scorpion to kill the hunter and, with its deadly stinger, the scorpion ended Orion's bragging.

As a reward, the scorpion was placed in the heavens. So was Orion, but to avoid any further battles between them, the two are never in the sky at the same time. We see Orion during the winter and Scorpius in the summer.

The stars of Orion look like a warrior to many people. But the Inuit who live in the far north saw a totally different scene. The bright star Betelgeuse was a bear. The three stars in Orion's belt were hunters, and the stars hanging from Orion's belt were the hunters' sled.

Look carefully at Betelgeuse, in Orion's right shoulder. Now look at Rigel, in his left knee. Do they look like they're the same color? With practice, you'll see that they're not. Betelgeuse is slightly redder than Rigel. Both are supergiant stars, much larger than our Sun. But, Rigel is much hotter than Betelgeuse—that's why the two stars are different colors.

Now look at Orion's belt. Do you see something hanging from it? That's an object called M42, the Orion Nebula. It is a huge cloud of gas in space, 1,500 light years away. Scientists think that the gases in a nebula like Orion are forming into stars.

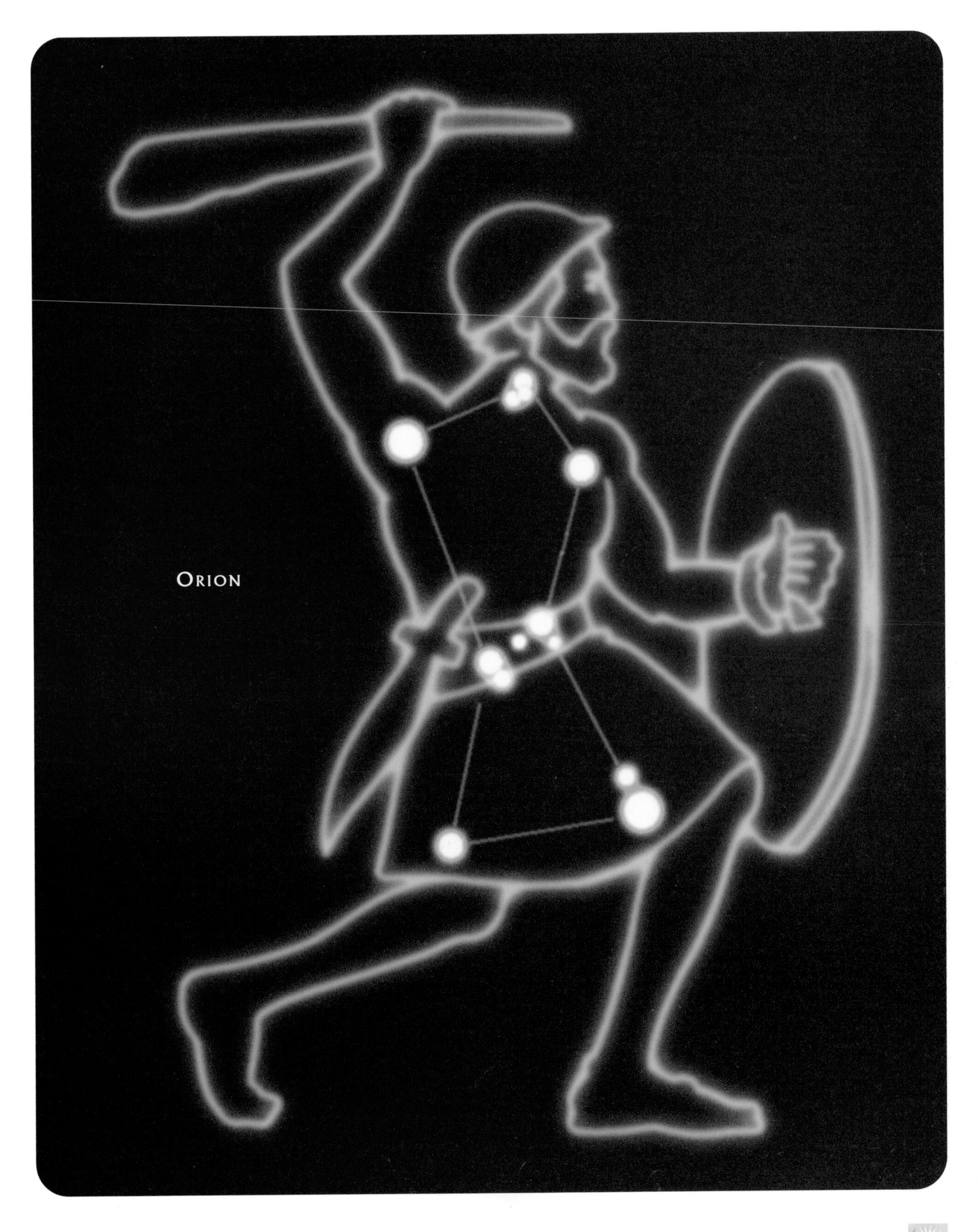
ORION

PAVO: The Peacock

THIS peacock struts his stuff near the south celestial pole. But it's nowhere near as proud as a regular peacock. It's not a large constellation and there isn't much to look at. It doesn't look like a peacock either.

It is one of the constellations invented by Keyser and de Houtman over 400 years ago.

PEGASUS: The Winged Horse

WITH its spreading wings and galloping legs, Pegasus is the seventh largest constellation in the sky. You can imagine a horse in these stars, as long as you imagine just the front half. Pegasus flies through autumn skies, nestled between the fish of Pisces and the Milky Way.

The body of Pegasus is also known as the Great Square. This asterism is sometimes seen as a baseball diamond by stargazing sports fans. Can you see a catcher and some outfielders?

According to Greek mythology, Bellerophon was the son of Poseidon. He was sent to kill the Chimaera, a strange fire-breathing beast that was part goat, part lion, and part dragon. Bellerophon would've had a hard time against the beast, but he had the help of Pegasus. Riding the winged horse, Bellerophon swooped down on the Chimaera and killed it.

When you look up at the stars, do you wonder whether there are planets out there beyond the planets in our solar system? Astronomers ask the same question. By looking at Pegasus, they seem to have finally found the answer.

Using very special equipment, astronomers have discovered a planet around a star called 51 Pegasi. This star is similar to the Sun, but the planet is very different from the Earth. It is closer to 51 Pegasi than Mercury is to the Sun, which means the planet must be very hot. And they think it is very large—like Jupiter or Saturn.

In fact, astronomers have been finding more and more planets in orbit around stars. Who knows—maybe someday, we'll discover life on one of those planets.

Pavo
Pegasus

PERSEUS: The Hero

PERSEUS is a fairly large constellation that can be seen during much of the year, but is highest in the late autumn night sky. Two lines of stars form a crooked V along the Milky Way, pointing toward Cassiopeia.

Perseus is the mythological Greek figure who rivals Hercules as the greatest hero in the night sky. On star maps, we usually see him carrying a sword in one hand, and a very unusual object in the other.

Hercules may have had Twelve Labors to perform, but Perseus had to kill a gorgon. Gorgons were monstrous creatures. Some say they had necks covered in dragon skin, tusks like a wild boar, hands of bronze, and wings of gold. Medusa was a gorgon whose head was covered in snakes. She was so hideous that if you took one look at her, you turned to stone. Talk about a bad hair day!

Perseus fought Medusa without looking directly at her. Instead, he looked at her reflection in his shield and was able to cut off her head. And that's what we see Perseus carrying in his hand—Medusa's snake-covered head!

The gorgon's head was more than just a great souvenir. It came in very handy during another of Perseus' adventures. As he was returning from his battle with Medusa, his journey carried him through King Cepheus and Queen Cassiopeia's kingdom. There he discovered their daughter, Andromeda, chained to the rocks as a sacrifice to Cetus the sea monster. Perseus fell in love with Andromeda on sight.

As Cetus approached the shore, Perseus held up Medusa's head so the monster could see it. The creature turned to stone and sank beneath the waves. Perseus released Andromeda from her chains and the two were married. And whenever anyone asked, "So, how did you two meet?" they always had a great story to tell.

Maybe the best meteor shower of the year comes from Perseus and is called the Perseid meteor shower. Go Perseid watching around August 12. You could see as many as one meteor a minute.

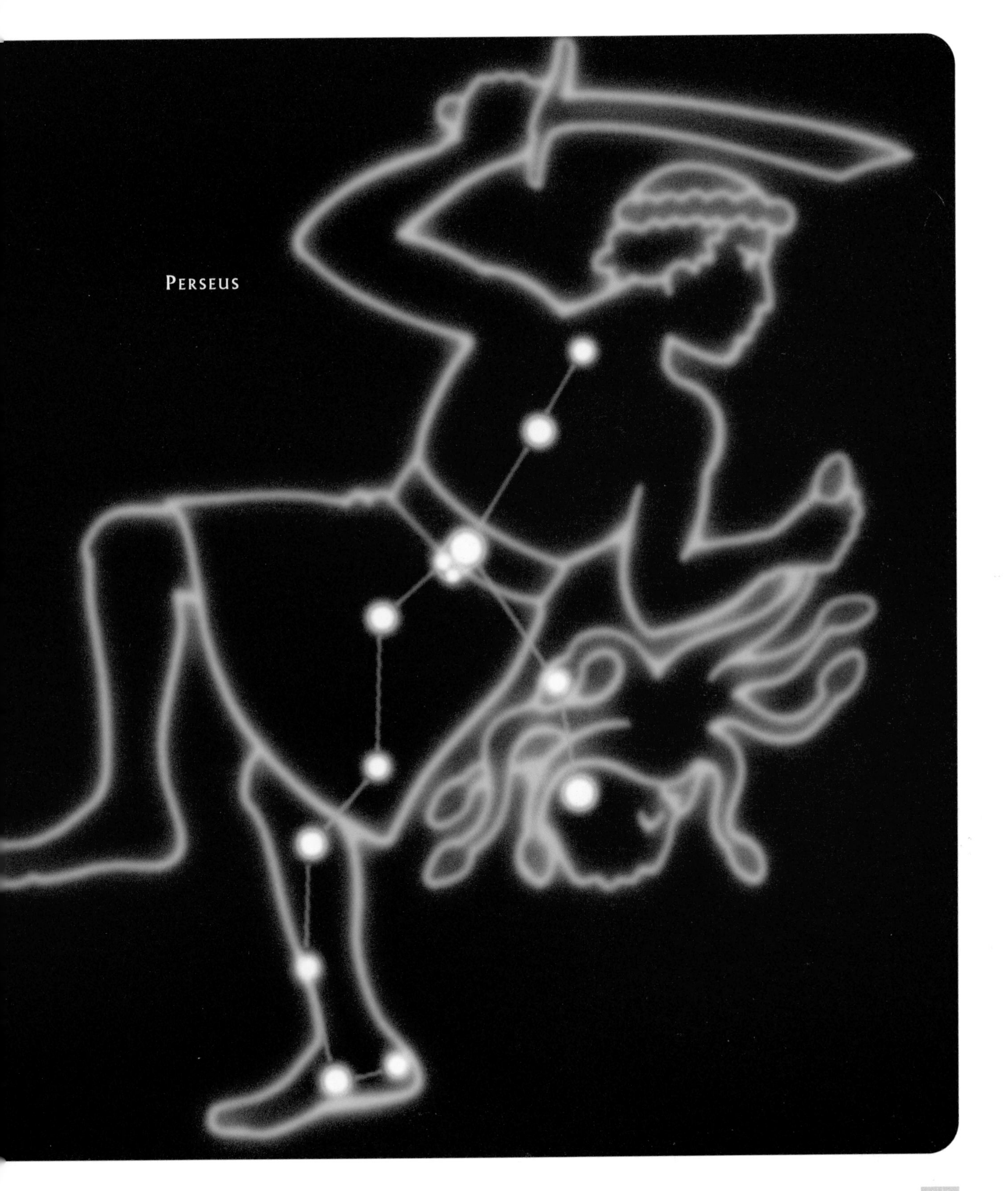
Perseus

PHOENIX: The Phoenix

PHOENIX is a medium-sized constellation, best seen in the southern spring sky. It flies through the sky near two other celestial birds, Grus the Crane and Tucana the Toucan.

It is one of the strange creatures placed in the sky over 400 years ago by Dutch navigators. The Phoenix is the mythical bird that is consumed by flame, but then reborn from the ashes of the fire.

PICTOR: The Painter's Easel

PICTOR is a small southern constellation that you can see best during the southern summer. But there isn't much to see. It's not big and doesn't have very bright stars.

It is one of de Lacaille's constellations named after the tools that scientists and artists use.

The second brightest star in Pictor is Beta Pictoris. Astronomers are interested in this star because it has an enormous disk of dust around it. They think that this disk of material will eventually turn into planets. But, don't hold your breath. If they're right, it will still take millions of years for the planets to form.

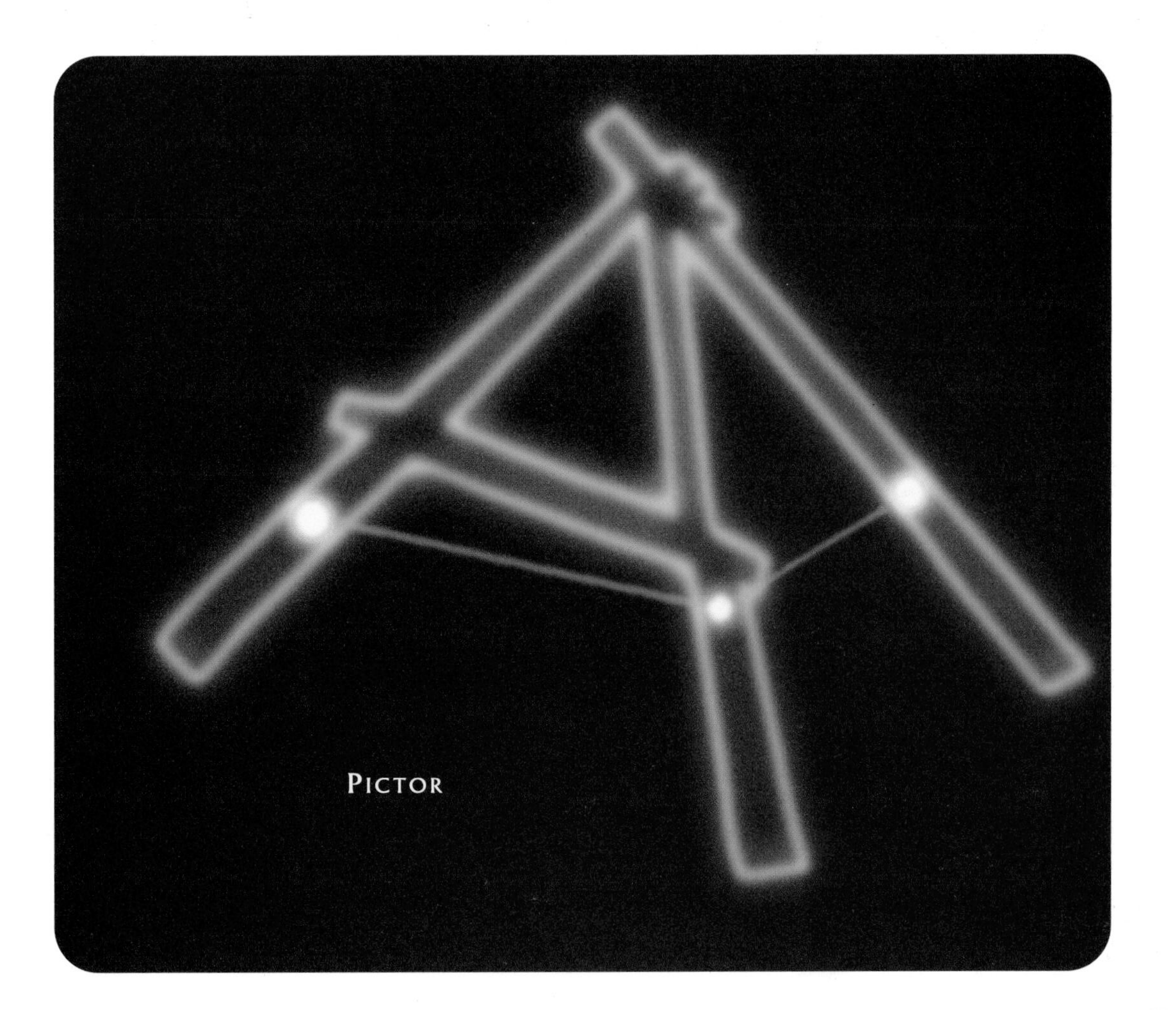

PISCES: The Fish(es)

PISCES is one of the constellations of the zodiac best seen in the autumn. It lies between its zodiac companions Aries and Aquarius. Two ribbons of stars form a V in the sky, with a pair of fish at the ends of the ribbons. The southernmost fish is marked by a grouping of stars that forms an asterism called the Circlet.

This part of the sky is sometimes called "the waters." It's where you'll find a sea monster, sea goat, dolphin, water bearer, and several fish.

To the Greeks and Romans, the celestial fish of Pisces were Aphrodite (Venus) and her son, Eros (Cupid). To escape the many-headed monster named Typhon, the two jumped into the Euphrates River. Once in the water, they turned into fish and swam away to safety. With their tails tied together, they would never be separated.

Did a very special star appear in Pisces, two thousand years ago? Back then, many people believed that things that happened in the sky reflected what happened on Earth. People who watched the heavens back then saw Jupiter and Saturn move through the sky and pass each other. This wasn't anything special. Planets are always passing close to each other in the sky. But then, the two planets passed each other again...and again. This was something that hardly ever happened. It's called a triple conjunction. Some people think the triple conjunction is what we now call the Star of Bethlehem. It wasn't a real star, just an event in the sky that people thought foretold the birth of the baby Jesus.

Does that mean the triple conjunction really predicted the birth of Christ? No. People often observed strange things in the skies like eclipses or novas, and thought they reflected an event on Earth or meant something important was going to happen. The triple conjunction was simply a rare event. And star watchers told a story about it foretelling the birth of a very special baby.

PISCIS AUSTRINUS: The Southern Fish

PISCIS AUSTRINUS is a fish in the part of the sky sometimes called "the waters." It's a small constellation that looks a bit like a fish. It even has a mouth—the brightest star in the constellation is named Fomalhaut, which is Arabic for "fish's mouth." Stargazers can catch Piscis Austrinus in autumn skies.

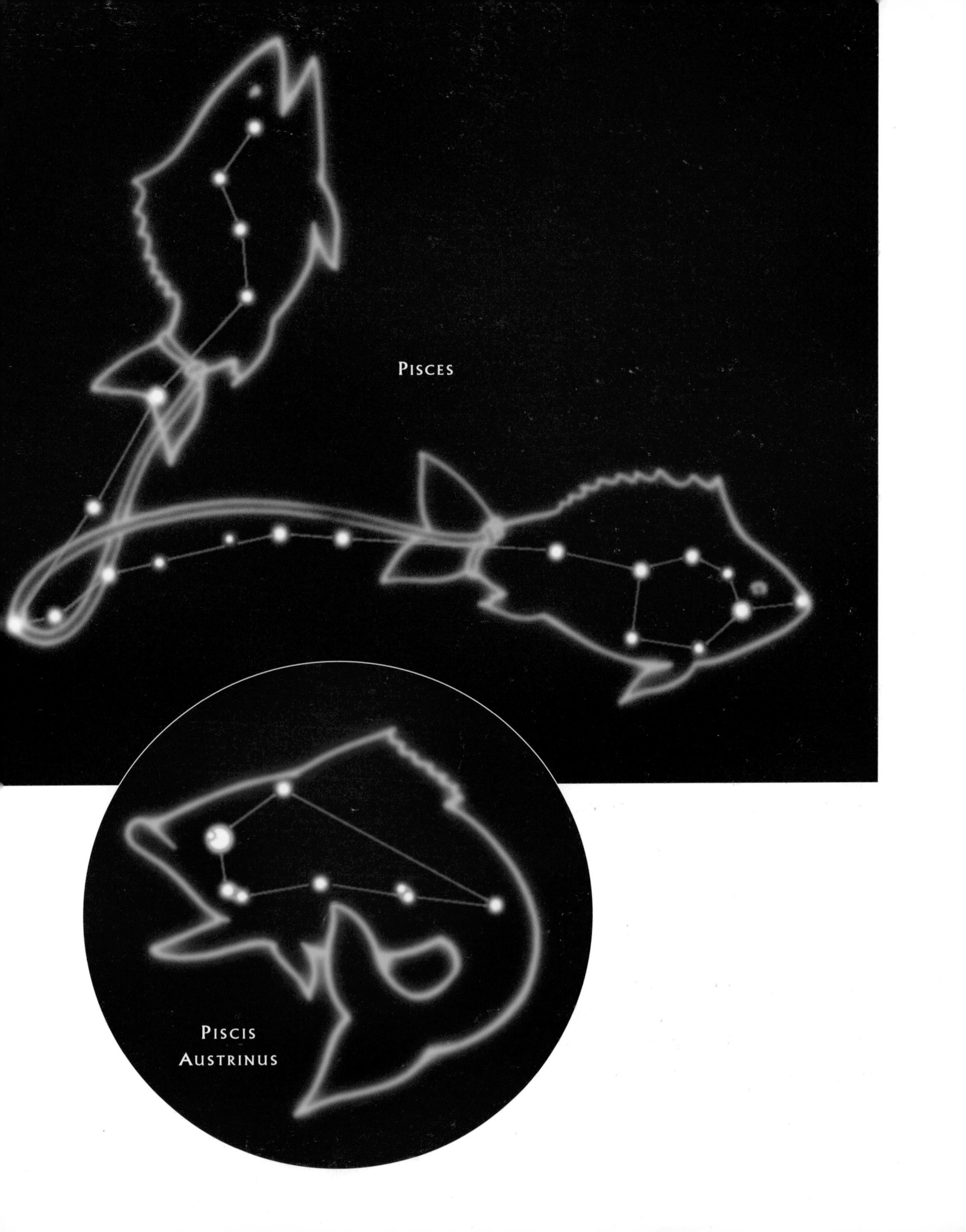
PISCES
PISCIS
AUSTRINUS

PUPPIS: The Stern (of *Argo Navis*)

PUPPIS is a large constellation, sailing through the Milky Way near Canis Major. It doesn't have any really bright stars and the stars don't make much of a pattern. But you might look for it anyway, when you're doing some winter stargazing.

Before Puppis appeared on star maps, there was a large constellation called Argo Navis. Jason and the Argonauts sailed the ship *Argo Navis* to many great adventures. But 250 years ago, the French astronomer Nicolas Louis de Lacaille divided this large constellation into three parts—Carina the Keel, Vela the Sail, and Puppis the Stern.

PYXIS: The Compass (of *Argo Navis*)

PYXIS is a small constellation in southern skies, with only a handful of visible stars that hardly looks like anything. You can see it best around February and March.

It is one of de Lacaille's constellations, invented in the 18th century. Pyxis is a compass—the type of compass sailors would use to tell direction. It sits near what used to be the constellation of Argo Navis—the ship of Jason and the Argonauts. That's why some people think of Pyxis as the compass used by these mythical seamen.

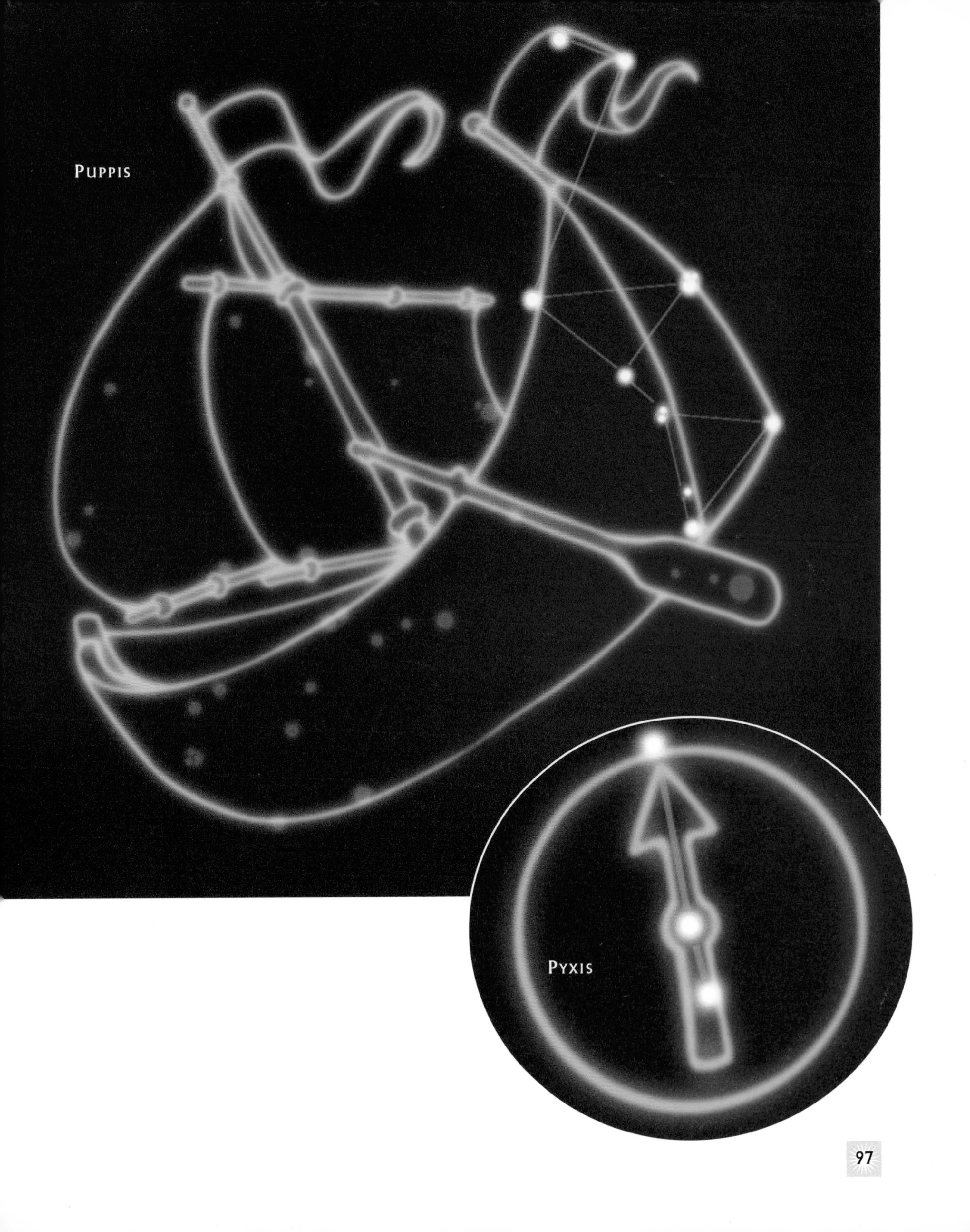
Puppis
Pyxis

RETICULUM: The Net

RETICULUM is another small constellation in the southern hemisphere. In fact, there are only six constellations that are smaller. You can find it on most nights, but it's highest in the sky in December.

This is one of the constellations invented nearly 250 years ago by de Lacaille. It represents something called a reticle—a part of a telescope that helps an astronomer aim it very accurately.

You won't see anything special within the constellation's boundaries. And because it's not very old, it isn't part of any Greek myth. But Reticulum still comes with a very interesting story.

A married couple named Betty and Barney Hill tell this tale. One night in September 1961 they were driving home when Betty noticed a strange light in the sky. They eventually made it home, but both Betty and Barney felt as if something strange had happened and couldn't remember what it was. It bothered them so much that they decided to visit a doctor. The doctor tried to find out what happened by hypnotizing them.

The Hills say they were then able to remember what happened that night. They describe the strange light as a flying saucer. They say that aliens—extraterrestrial creatures from another planet—took them into their spaceship. Betty even says she asked the aliens where they were from and that they showed her a map of their home star. According to some people, the map that Betty drew from memory shows that the aliens came from a planet in orbit around a star in Reticulum, called Zeta Reticuli.

Were Betty and Barney Hill really abducted by aliens? Nearly all scientists will tell you there's no scientific evidence that extraterrestrials exist. They'll also tell you that hypnosis doesn't necessarily help you remember things. Maybe the Hills only think these things happened. Maybe they made up their story. Maybe, like the other stories that people tell about the stars—of dragons, sea monsters, and warriors—these stories are just tales we like to tell about ourselves and the world around us.

RETICULUM

SAGITTA: The Arrow

SAGITTA is the third smallest constellation. It doesn't have many stars, but they look a bit like an arrow. On summer nights, you'll see it shooting through the Milky Way.

Through time, people thought of this constellation as many different arrows. Some thought of it as Cupid's arrow, some as the arrow that Apollo used to kill the Cyclops. Others said it was one of the arrows that Hercules used to hunt the Stymphalian Birds, one of his Twelve Labors.

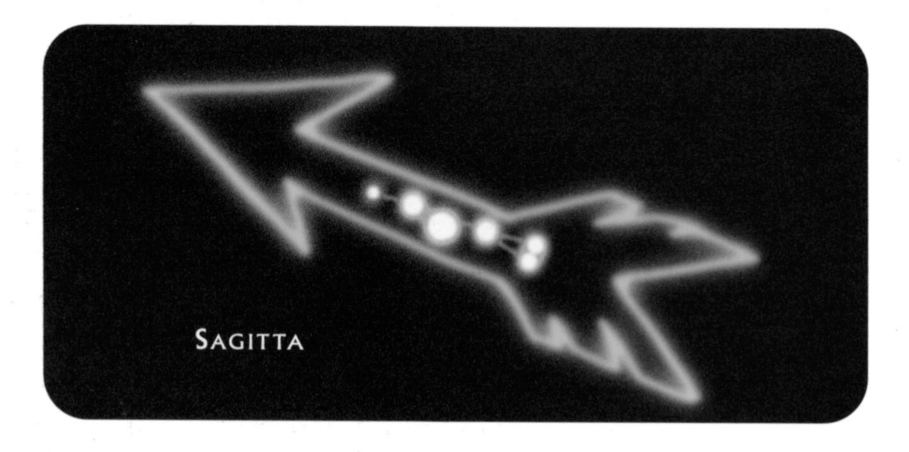

SAGITTARIUS: The Archer

SAGITTARIUS is one of the constellations of the zodiac, lying between Scorpius and Capricornus. It straddles the Milky Way in summer skies.

Eight of its brighter stars form a pattern that looks a lot like a teapot. The stars of nearby Corona Australis look like a slice of lemon. You could even imagine that the Milky Way is steam rising from the spout of the teapot.

Different people have seen many different archers in the stars of Sagittarius. To the Sumerians, he was their god of war. To the Greeks, he was Crotus, the inventor of archery, aiming his arrow at Scorpius.

Find Sagittarius in the night sky. You're looking toward the center of the Milky Way galaxy.

Our galaxy is a disk of hundreds of millions of stars, a hundred million light-years in diameter. Imagine two dinner plates placed together. That's the shape of our galaxy. Now imagine two arms spiraling out from the center, and you have a good idea what our galaxy looks like.

Our Sun is nowhere near the heart of the Milky Way Galaxy. In fact, it's closer to the edge of the galaxy than the center. But you can still find the center, just by looking at Sagittarius.

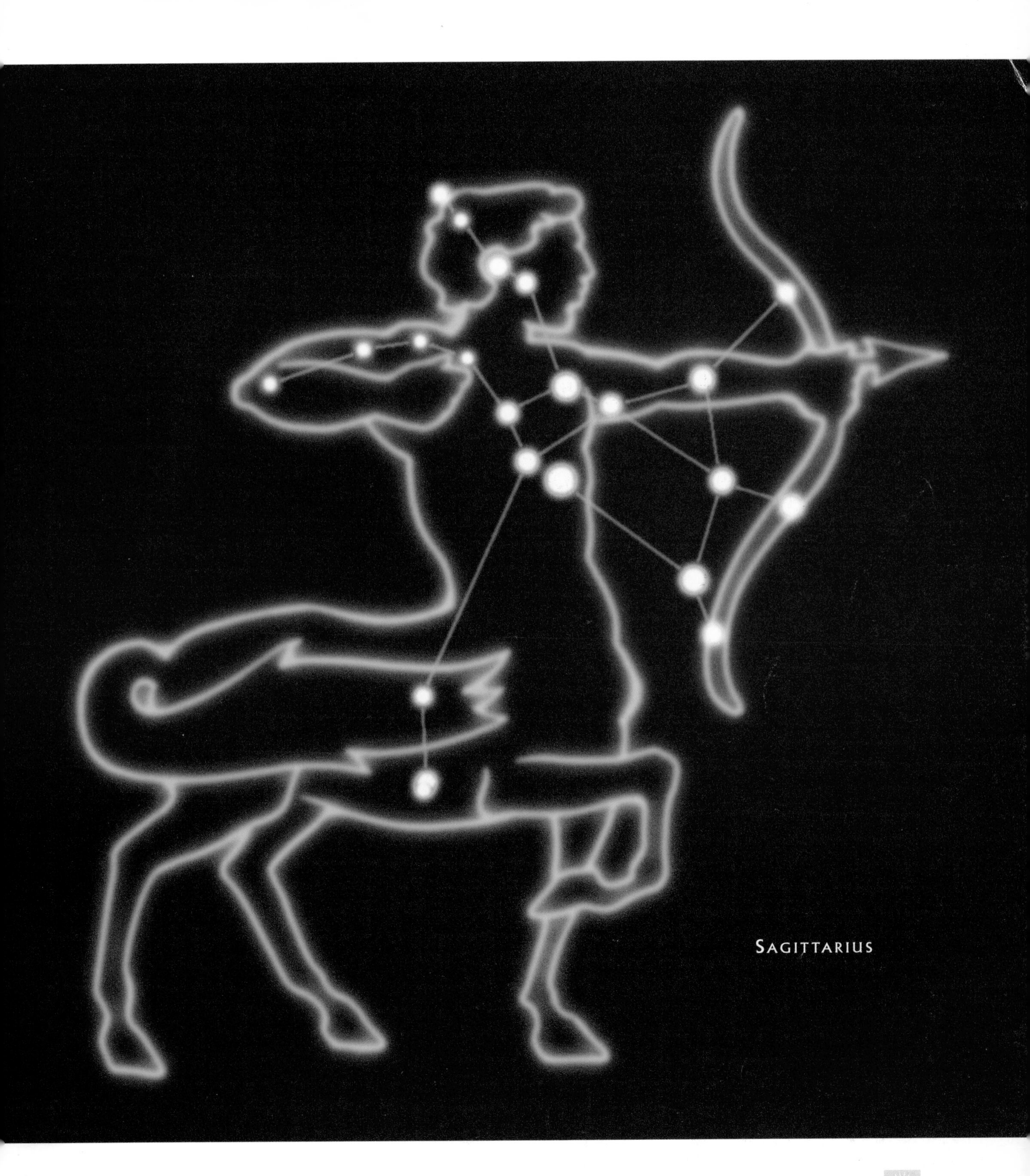
SAGITTARIUS

SCORPIUS: The Scorpion

IT'S not hard to see the scorpion that this constellation of the zodiac was named after. Stars form the curving body of the scorpion and end in the curl of its poisonous tail. There are even claws if you add the stars of Libra, as ancient sky watchers did.

Most star watchers see Scorpius crawl along the southern horizon during the summer, between Sagittarius and Libra.

In Greek mythology, Scorpius was the creature who killed Orion. To the Maori of New Zealand, the stars that make the scorpion's tail are the fish hook that was thrown into the ocean by the god Maui. When the hook was retrieved, it pulled an island up from the ocean. Maui's fishermen cut the island in two and they became the islands of New Zealand.

The heart of the scorpion is marked by a very bright star called Antares, which means "rival of Mars." It is a red supergiant star, hundreds of times bigger than the Sun. Look at Antares. Can you see that it is slightly redder than most other stars? It looks like Mars, which also shines with a reddish glow.

SCULPTOR: The Sculptor's Workshop

THIS small, faint constellation has few stars. So don't be surprised if you can't imagine a sculptor's table in the sky. It's only on star maps that you'll see Sculptor as a marble bust, sitting on a table waiting for the artist.

It is one of the constellations invented by de Lacaille.

SCUTUM: The Shield

SCUTUM is the fifth smallest constellation in the sky, with few stars. It hangs in the Milky Way, near Aquila and Sagittarius. It doesn't have many stars, but it does look something like a shield.

This constellation was invented by the Polish astronomer Johannes Hevelius. The shield actually belongs to someone. Hevelius first called the star grouping "Sobieski's Shield," in honor of King John III Sobieski of Poland.

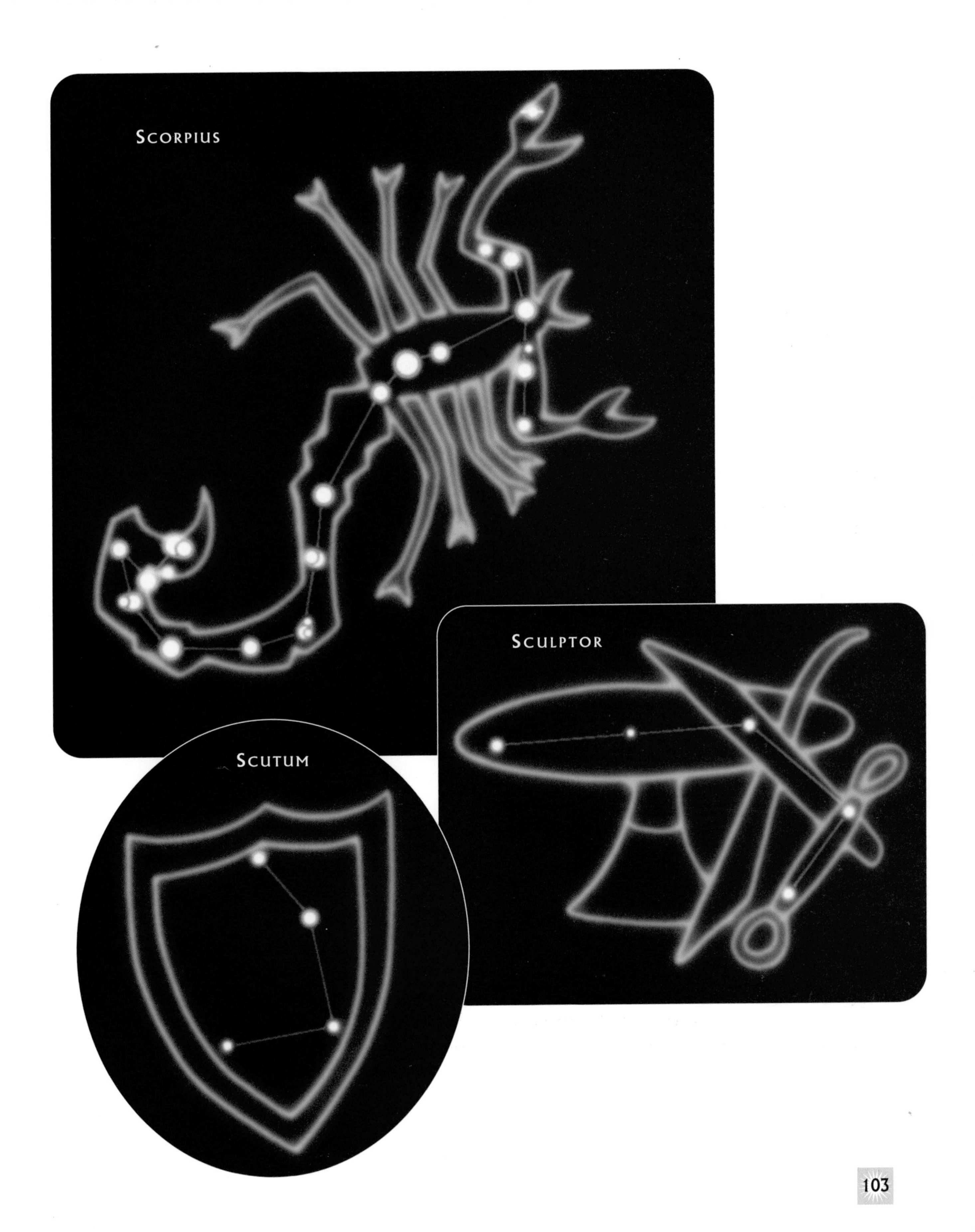

Scorpius
Sculptor
Scutum

SERPENS CAPUT/CAUDA: Head/Tail of the Snake

SERPENS is unlike any other constellation. It is the only constellation that is made up of two different areas of the sky that aren't connected. Both parts look like trails of stars that together form a snake writhing across summer skies.

The two halves of the serpent are separated by Ophiuchus, the Serpent Bearer. On star maps, you can see him holding Serpens Caput with one hand and Serpens Cauda with the other. The head hisses at Boötes and Corona Borealis, while the tail slithers near Aquila.

The Eagle Nebula is a vast cloud of gas and dust, 7,000 light-years away in Serpens. Not long ago, astronomers used the Hubble Space Telescope, in orbit around the Earth, to photograph it. In the photograph you'll see enormous fingers of gas and dust in the nebula. How big are the fingers? They're about a light-year tall. That means a beam of light would take about a year to go from the bottom of the photograph to the top.

Deep inside the fingers of the nebula, clouds of gas and dust are shrinking. As they shrink, they'll get hotter and hotter. After millions of years, they'll finally be hot enough to turn into stars.

SEXTANS: The Sextant

SEXTANS sits between Hydra and Leo in spring skies, but there's not much to look at. There aren't many stars and they aren't very bright.

But, you can understand why the Polish astronomer Johannes Hevelius would name a constellation after a sextant. He probably spent a lot of time using one to accurately measure the positions of stars for his star atlas.

Serpens Caput/Cauda
Sextans

TAURUS, the Bull, is a constellation of the zodiac. The bull, charging across the winter night sky, is one of the best star pictures in the sky. The bull's face is outlined by a V-shaped cluster of stars, the reddish star Aldebaran is the bull's bloodshot eye, and its long horns extend upward to two bright stars. Taurus is glaring menacingly at the hunter Orion, as if the two are trying to stare each other down in a celestial bullfight.

The bull's eye, Aldebaran, is also one of the bright stars in the "Winter Circle" or "Winter Hexagon."

Sitting on the bull's shoulder is a cluster of stars called the Pleiades. The cluster is one of the most beautiful sights in the night sky, whether you're looking at them with the naked eye or with a large telescope. People around the world tell many stories about the Pleiades.

To the Greeks, they were seven stars who represented the beautiful daughters of Pleione and Atlas. One day, while traveling with their mother, the seven sisters met the hunter Orion. They were so beautiful that he fell in love with all of them immediately, and began to chase them. But, Zeus didn't approve and ended the chase by changing the sisters into doves. They flew into the sky, where we still see them today. And Orion? He hasn't given up. Every winter, you can see him chasing the sisters across the night sky.

To the Iroquois, the seven stars of the Pleiades were seven young boys who, eager to grow into warriors, danced a war dance. But, they were so enthusiastic that their dancing lifted them right off the ground and into the sky. As they rose, one of the boys stopped dancing and fell back to earth, leaving six stars in the sky. And, in fact, most people see only six stars when they look at the Pleiades.

Do you think you've seen these six stars somewhere else? Maybe you've seen them while stuck in traffic. In Japan, the six stars are called Subaru. You may have seen them on the cars of the same name.

The V-shaped cluster of stars that make up the bull's face are called the Hyades. They and the Pleiades are actually clusters of stars, held together in space by gravity. There are about 200 stars in the Hyades, all about 150 light-years from us. There are actually hundreds of stars and great clouds of gas that make the Pleiades one of the most beautiful star clusters that astronomers can photograph.

For years, scientists have been listening to a very faint radio signal coming from the direction of Taurus. Is it an alien civilization trying to contact us? No, it's actually a very faint radio signal from a spacecraft named Pioneer 10. The space probe was launched in 1972 and was the first to fly by the planet Jupiter. Eventually, Pioneer 10 left the solar system and has been heading in the direction of Taurus ever since. Today, it's over 11 billion kilometers from Earth. Just in case any extraterrestrial creatures find this interstellar spaceship, Pioneer 10 carries a picture showing what humans look like. Imagine what aliens will think of these strange creatures with two arms, two legs, and only one head!

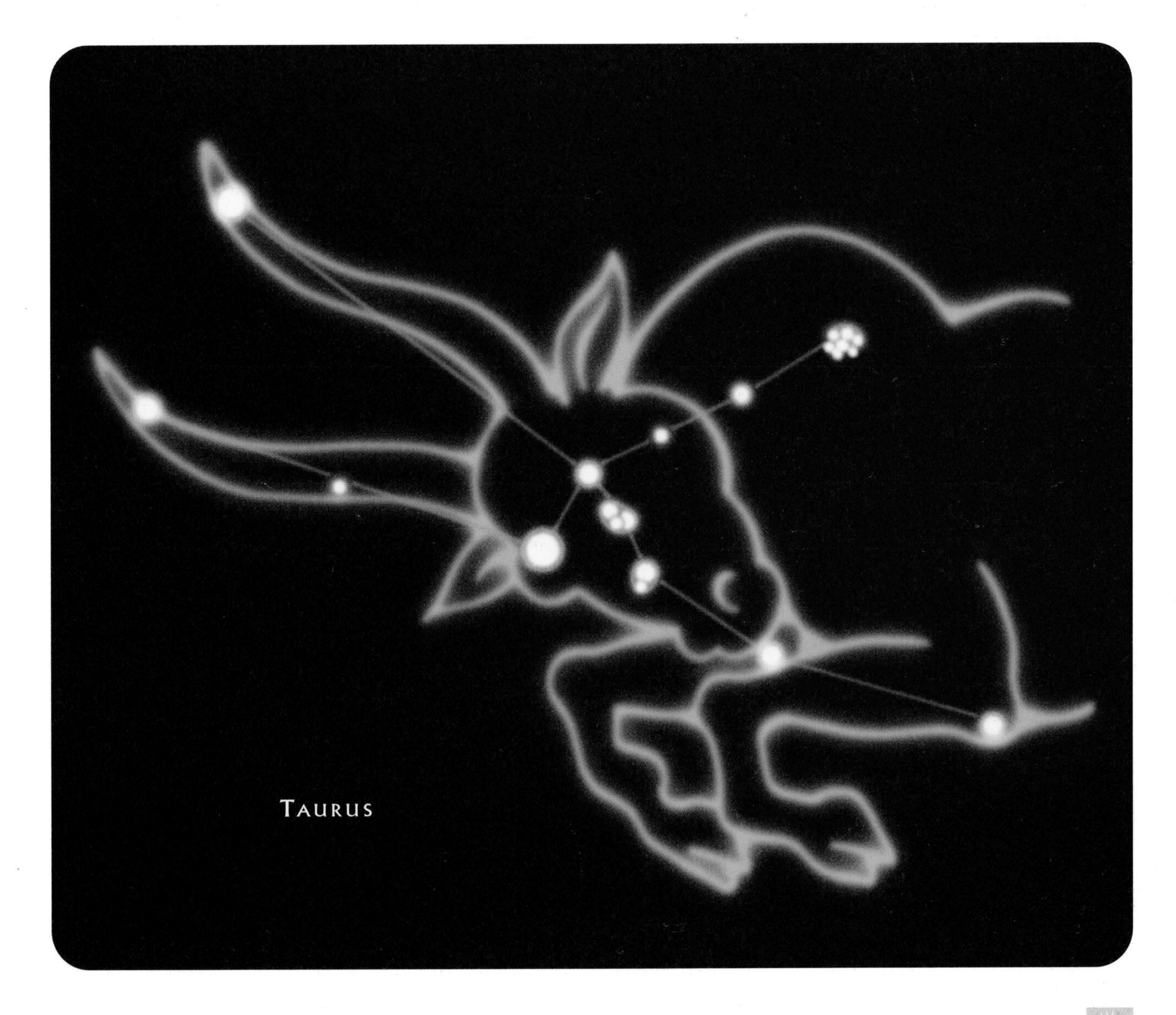

TELESCOPIUM: The Telescope

TELESCOPIUM is hard for most northern observers to see. It's too small, it doesn't contain any bright stars, and it's too close to the horizon.

It's one of the constellations named by de Lacaille. We shouldn't be surprised that there is a group of stars honoring the telescope. After all, when we started looking at the night sky with a telescope, it showed us what the universe was really like. We saw that other planets had moons, the Sun had sunspots, our Moon was covered with craters and mountains—and we saw that there were many, many, many more stars than we imagined!

Telescopes work because they make distant objects seem closer, and because they also make it easier for us to see very faint objects—by gathering in much more light than our eyes can. Light enters our eyes through the pupil—that's the black dot in the middle of your eye. Compare the size of your pupil to the lens of even a small telescope and you'll get an idea of how much more light a telescope can collect than your eye.

Astronomers build very large telescopes on mountaintops. That way, they don't have to look through miles and miles of air. Even though we don't think about it, looking through air makes stars blurry. By building telescopes on mountains, astronomers look at the stars through as little air as possible.

That's why scientists built the Hubble Space Telescope, a large telescope in orbit around the Earth. Because it's in space, it doesn't have to peer through any air at all. That's one reason why photographs taken through the space telescope are so amazing.

Little did de Lacaille imagine that his imaginary telescope in the sky would one day be joined by a real one!

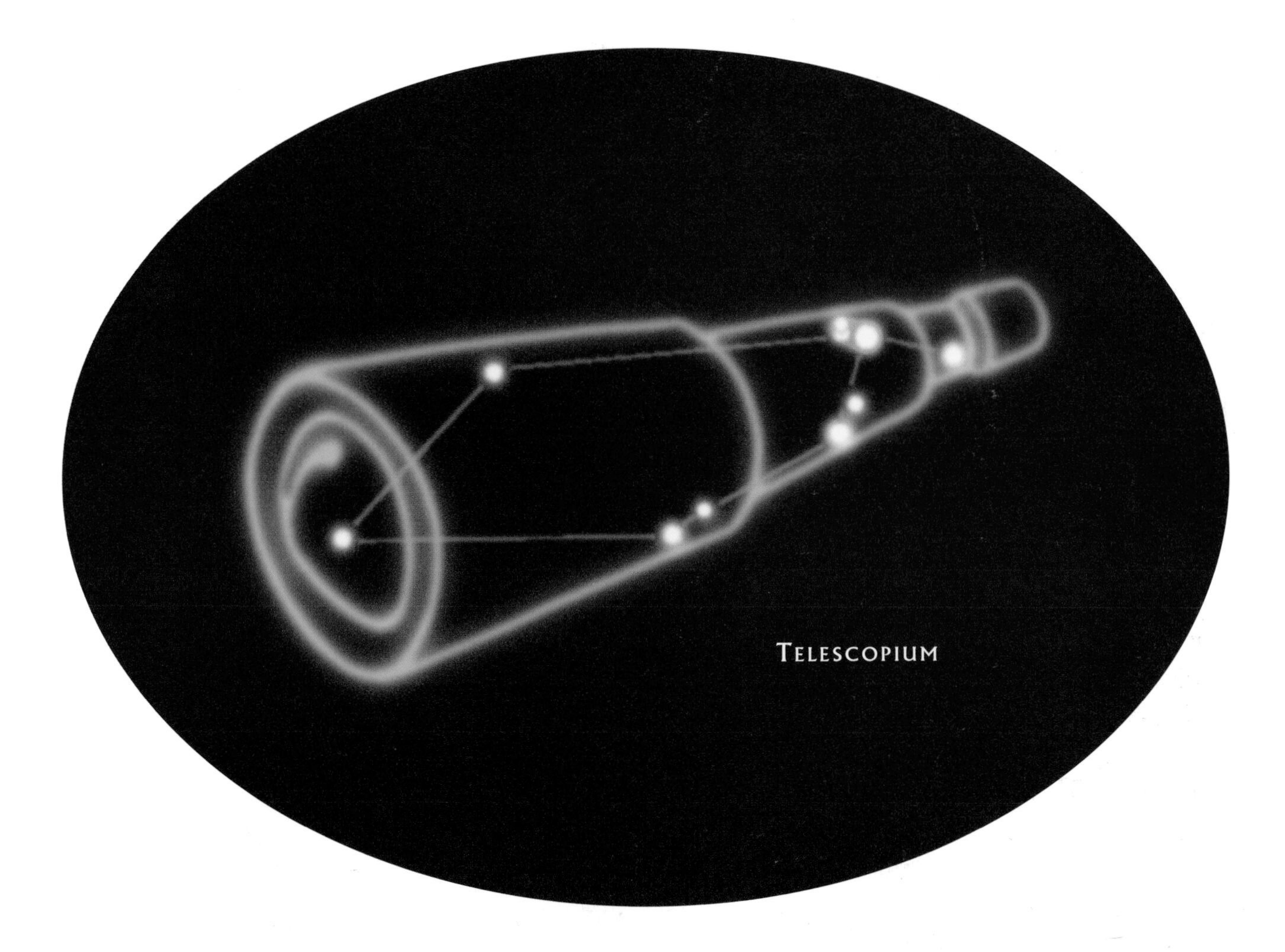
TELESCOPIUM

TRIANGULUM: The Triangle

GUESS what the constellation Triangulum looks like. Of course, Triangulum looks like a triangle. It's a very small autumn constellation, with no bright stars, wedged between Pisces, Aries, Perseus, and Andromeda.

To the ancient Greeks, this triangle of stars looked like the delta of the Nile River. To others, it looked like the island of Sicily.

Triangulum may be small, but it contains an interesting object. M33 is a spiral galaxy that is like our galaxy and the Andromeda Galaxy. It's over 2 million light-years away and harder to see than the Andromeda Galaxy. But with some practice and a very dark, clear autumn night, you'll probably be able to see it.

TRIANGULUM

TRIANGULUM AUSTRALE: The Southern Triangle

TRIANGULUM AUSTRALE may be a little smaller than Triangulum, but its stars are brighter. It can be found in southern winter skies, not far from Alpha and Beta Centauri.

It's the smallest of the constellations named by the Dutch navigators Keyser and de Houtman at the end of the 16th century.

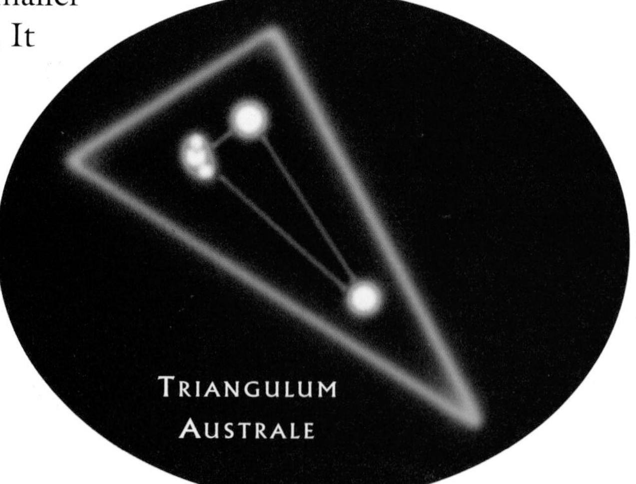

TUCANA: The Toucan

TUCANA is part of a flock of birds in the southern sky, along with Grus and Phoenix. Tucana isn't far from the south celestial pole, so you can see it most nights of the year. But it's highest in the skies in the southern spring.

The Toucan is another unusual creature placed in the sky by the Dutch navigators Pieter Dirkszoon Keyser and Frederick de Houtman over 400 years ago.

The Small Magellanic Cloud, which is found in Tucana, looks like a piece that broke off the Milky Way. The Large and Small Magellanic Clouds are actually small galaxies. They're the closest galaxies to our own, less than 200 thousand light-years away.

URSA MAJOR: The Great Bear

URSA MAJOR is one of the best known constellations in the sky. Most stargazers will find it in the northern part of the sky, no matter what season or time of night it is. It's the third largest constellation, and it has many bright stars. It even looks like a bear!

Within the Great Bear is the most famous asterism, the Big Dipper—a pattern of seven stars that look like a dipper or a pot with a long handle.

The Big Dipper is a great place to start when you're looking for other stars or constellations. Imagine a line through the two bright stars in the bowl of the dipper. That line points right to the North Star in Ursa Minor.

Or draw a curved line along the handle of the dipper. That leads you to a very bright star called Arcturus in Boötes. Or poke a hole in the bowl of the dipper. Whatever is in the dipper will pour out onto Leo. Or draw a line from the bend in the handle of the dipper to the North Star, keep going, and you'll come to Cassiopeia.

How did the bear get up in the sky? According to the ancient Greeks, the king of the gods, Zeus, fell in love with a huntress named Callisto. Zeus was so in love that he disguised himself as the god Artemis to gain her affection. Soon, Callisto gave birth to a son, whom she named Arcas. Zeus's wife wasn't too happy about this whole thing and, in a fit of anger, turned Callisto into a bear. But the story doesn't end there. One day, while hunting in the woods, Arcas came upon the bear. Without knowing it, he began to stalk his own mother. Seeing what was about to happen, Zeus sent Callisto out of the reach of her son's arrows, to the safety of the night sky. Today, we see Callisto as Ursa Major.

To some people, the Big Dipper is known as the Plough or the Wagon. The Inuit, who live in the Arctic, call it Tukturjuit or Tukturjuk, meaning "caribou." The Ojibwas of North America see it as a mink-like creature who was chased into the sky. The Micmac, who live on the east coast of Canada, see a large bear pursued by hunters.

To slaves in America in the 1800s, the Big Dipper was more than a pattern of stars—it meant freedom. They called it the Drinking Gourd and sang a song that told how to make your way north to the Underground Railroad. The Underground Railroad wasn't a real railroad. It was a network of people who helped thousands of slaves to escape slavery.

One verse went like this:

> *When the sun comes back and the first quail calls,*
> *Follow the Drinking Gourd.*
> *For the old man is waiting for to carry you to freedom,*
> *If you follow the Drinking Gourd.*

We still tell stories about the Big Dipper today. When the space shuttle *Challenger* exploded and seven astronauts lost their lives, astronauts on the next shuttle mission wore a special patch on their spacesuits. It featured the Big Dipper, with the seven stars representing the seven *Challenger* astronauts. So, depending on who you are, the stars of Ursa Major and the Big Dipper mean many different things.

STAND on the North Pole and look straight up, and you'll see Ursa Minor. It takes some imagination to see a bear in these stars, but try looking for a little bear cub with a very long tail. The long tail is the handle in the asterism of the Little Dipper.

The brightest star of this constellation is at the end of the dipper's handle. It's probably the most famous star of all—Polaris, the North Star. It's not the brightest star in the sky (in fact, in brightness it's #48), it's not the biggest, and it's not the closest. It's famous because it marks the north celestial pole—the point directly over the North Pole.

To the Inuit, Polaris is known as Nuuttuittuq, which means "never moves." In the play, *Julius Caesar*, William Shakespeare wrote that when Caesar would not change his mind, he was being "constant as the Northern Star."

Why doesn't the North Star move? Watch the sky for a few hours. You'll notice the stars move across the sky the same way the Sun and Moon move through the sky. We see this because the Earth is rotating, spinning around like a top, once every 24 hours. Watch carefully all night long, and you'll see that Polaris is the only star that doesn't move across the sky. It just sits in the north, always in the same spot, watching all the other stars move around it.

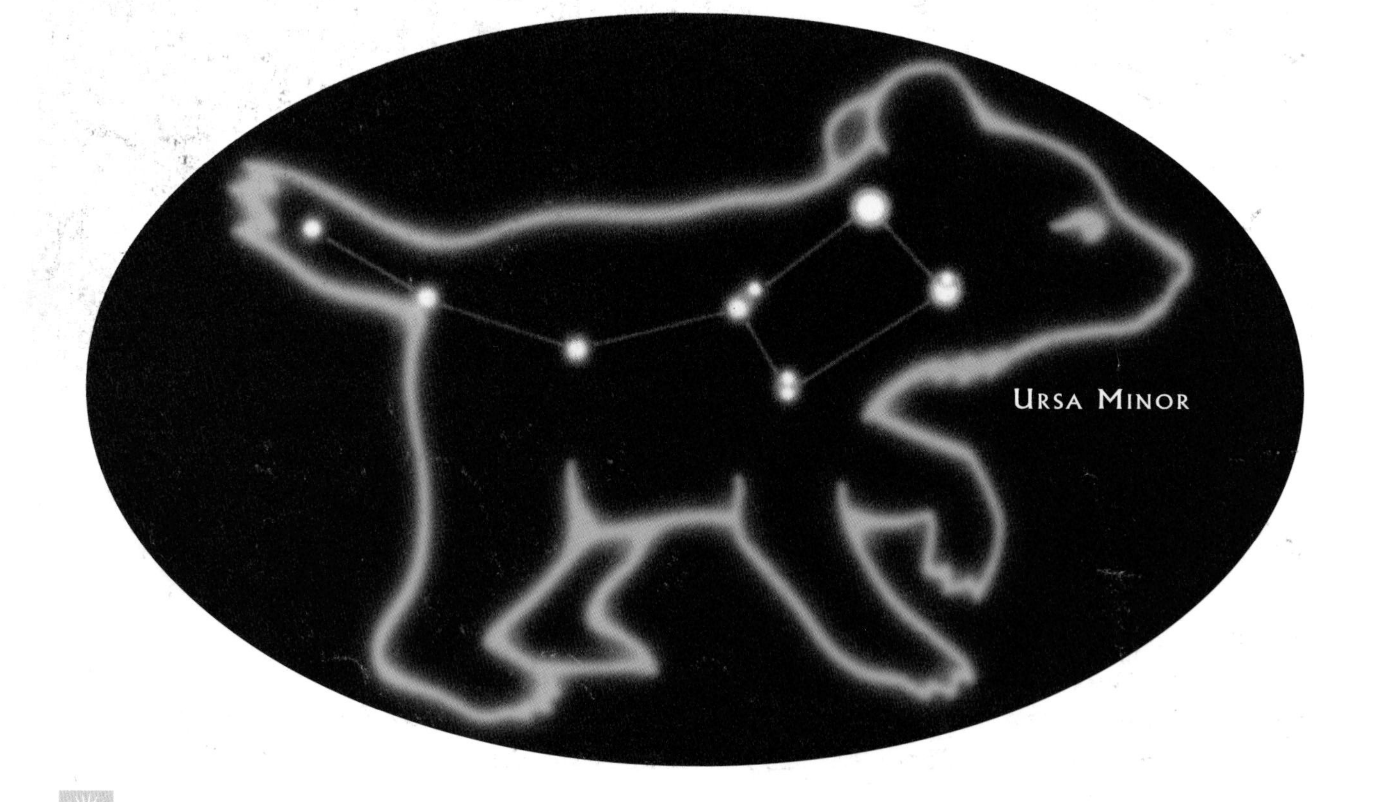

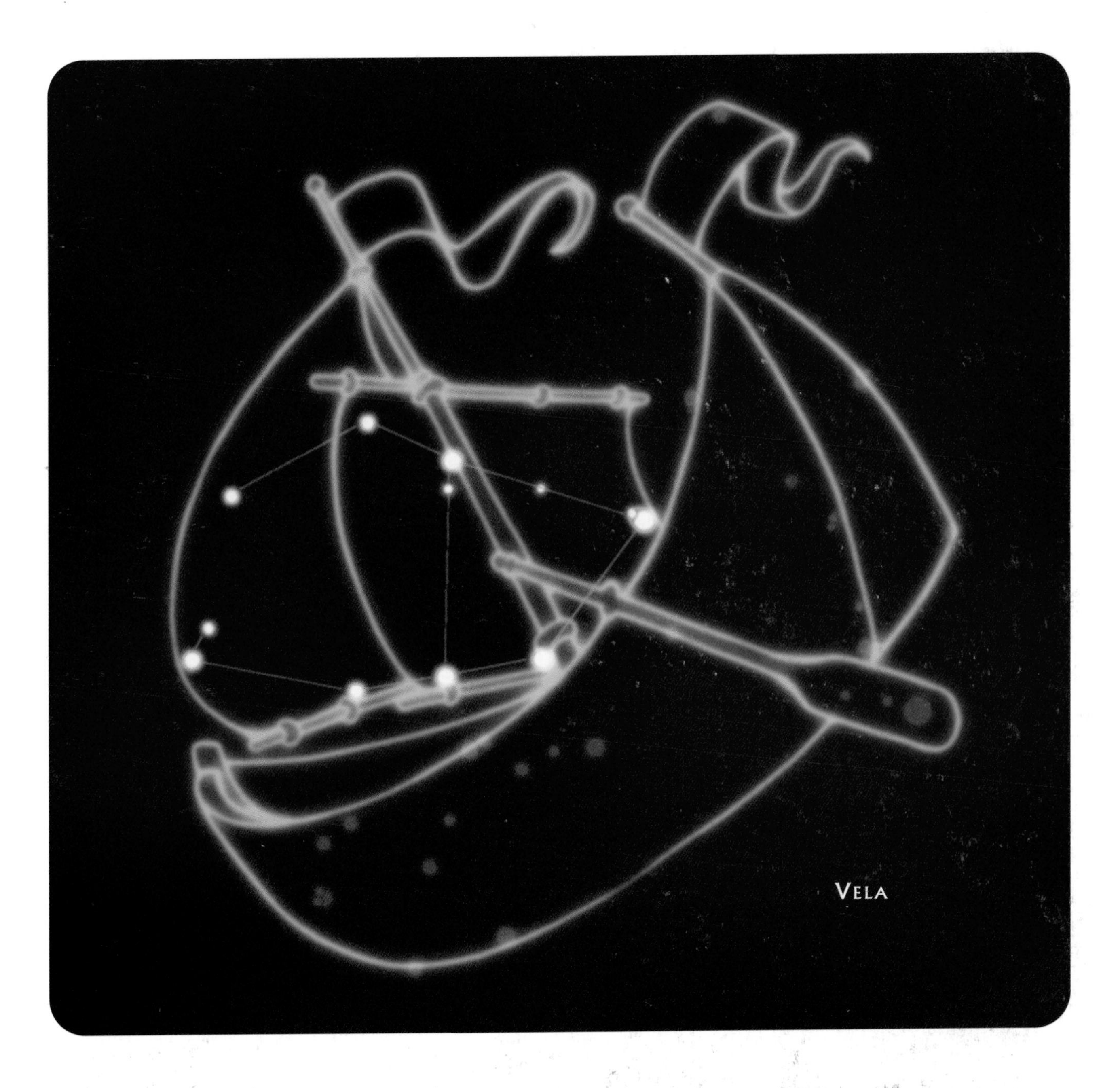

VELA: The Sail (of *Argo Navis*)

VELA is a fairly large constellation that you can see in the late summer skies of the Southern Hemisphere. It is the sail of the *Argo Navis*, the ship that Jason and the Argonauts sailed on many adventures.

Vela was once part of the larger constellation Argo Navis until some 250 years ago. That's when the French astronomer de Lacaille divided up Argo Navis up into three different constellations—Carina the Keel, Puppis the Stern, and Vela the Sail.

VIRGO: The Maiden, The Virgin

VIRGO the Maiden is the second largest constellation in the night sky, and the largest of the twelve constellations of the zodiac. We see Virgo stretching across spring skies, between Libra and Leo.

The bright star Spica is Latin for "ear of wheat." and that's usually what we see Virgo holding in her left hand. Spica gives us an easy way to find Virgo. Start with the Big Dipper, and draw an arcing line along the dipper's handle. That line takes you to Arcturus in Boötes. Now, quickly draw that same line beyond Arcturus and you'll come to Spica. A good trick for remembering how to find these two stars and constellations is the phrase "Arc to Arcturus...Speed to Spica."

Maybe because there aren't as many women in the sky as men, people saw many different female figures in the stars of Virgo. At different times and for different people she was Ishtar, Isis, Demeter, Athena, and Artemis. In India, she was the mother of Krishna. And to medieval Christians, she was the Virgin Mary.

Galaxies gather together in clusters. Our Milky Way Galaxy, the Andromeda Galaxy, the Large and Small Magellanic, M33, and a number of other galaxies form a group called the Local Group. But there are many other clusters. With a large enough telescope, we can see the Virgo Cluster sitting on Virgo's right shoulder. This is a collection of thousands of galaxies, about 50 million light years away.

You'd think a cluster like the Virgo Cluster might be the biggest thing in the universe. But there are also clusters of clusters of galaxies—called super clusters—that are even bigger. The universe is a big place!

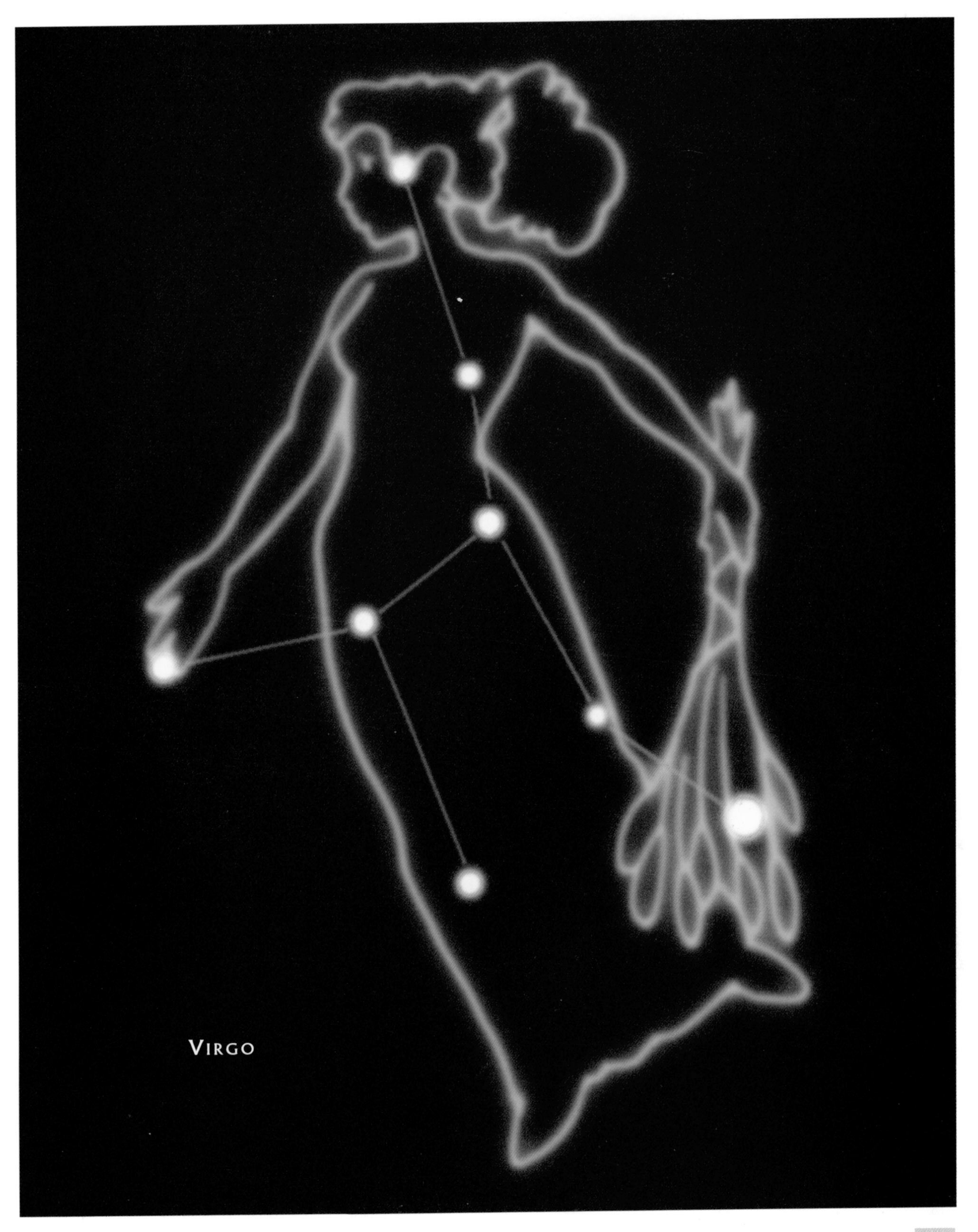
VIRGO

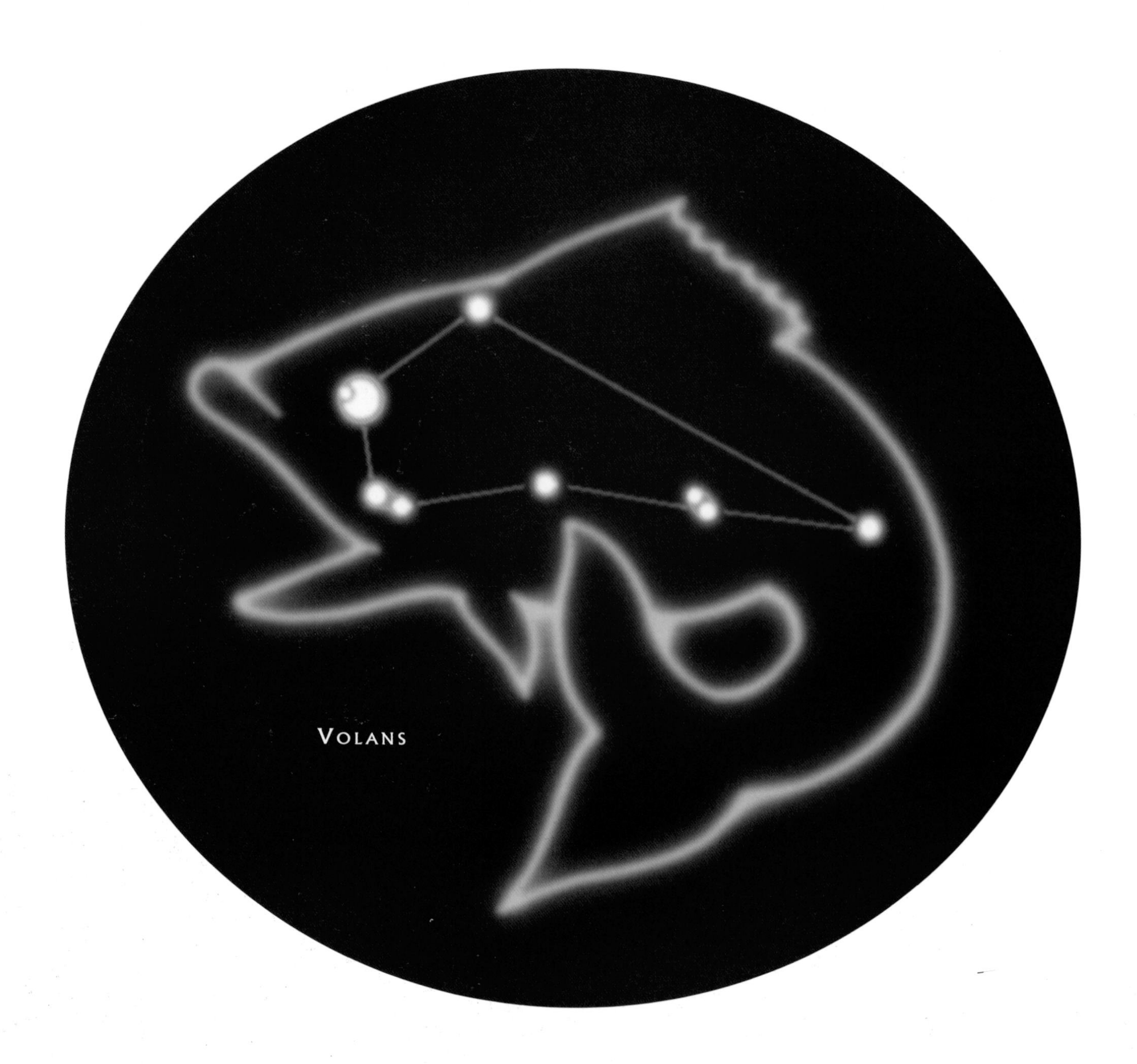

VOLANS: The Flying Fish

VOLANS the Flying Fish looks like it's leaping from the waters of the Milky Way. It's a small constellation in southern skies, with no bright stars, lying near the Large Magellanic Cloud. It's visible most nights of the year, but is best seen in January and February.

A constellation named after an exotic creature like a winged fish can only be one of the constellations invented by Keyser and de Houtman.

VULPECULA: The Fox

VULPECULA skulks through the Milky Way, between Cygnus and Delphinus, during summer nights. It's not a large constellation, and it has no very bright stars. It doesn't look much like a fox, either.

Johannes Hevelius named the Fox in 1690. At first, he called it Vulpecula cum Anser. That means "the Fox with the Goose." But over the years, the goose part has been forgotten as if it has been eaten by the fox or has flown away.

You can't see it, but there's a strange object in Vulpecula with an even stranger name. It's called PSR 1919+21. In 1967, an astronomer named Jocelyn Bell heard something unusual coming from this object. It was a radio signal, pulsing every 1½ seconds. No one knew what it was. Some scientists thought it might be a signal from aliens, so they called the object "LGM," meaning "Little Green Men." Eventually, astronomers figured out that the signal was coming from a special type of star that came to be known as a pulsar. Pulsars rotate much faster than our Sun or Earth. PSR 1919+21 rotates once every couple of seconds! With every rotation, a hot spot on the star sends a signal our way. It's like a stellar lighthouse that turns very, very quickly.

Since the discovery of the first pulsar in Vulpecula, astronomers have discovered many more. But so far, no little green men!

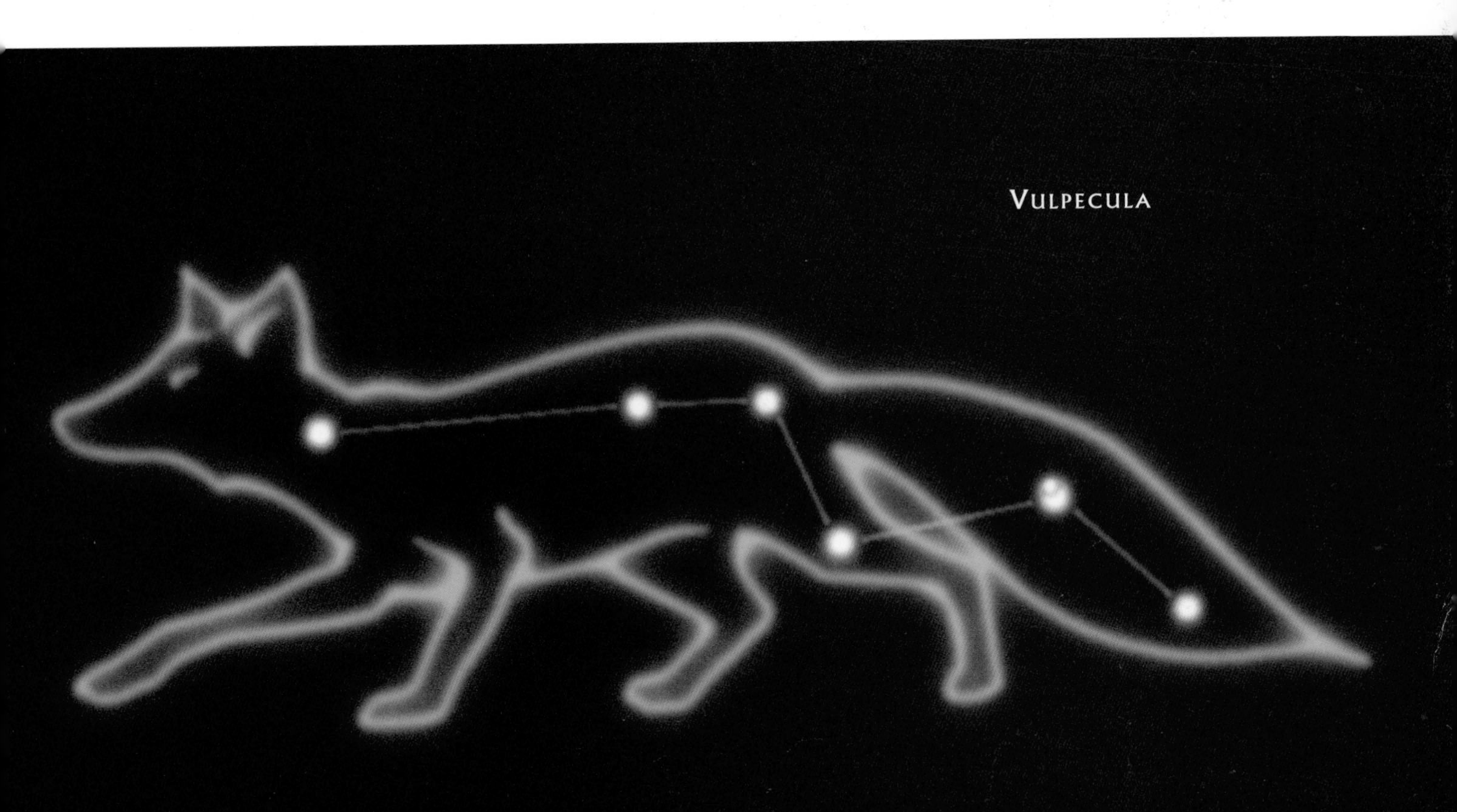

THE MAPS

MAPS FOR THE NORTHERN LATITUDES

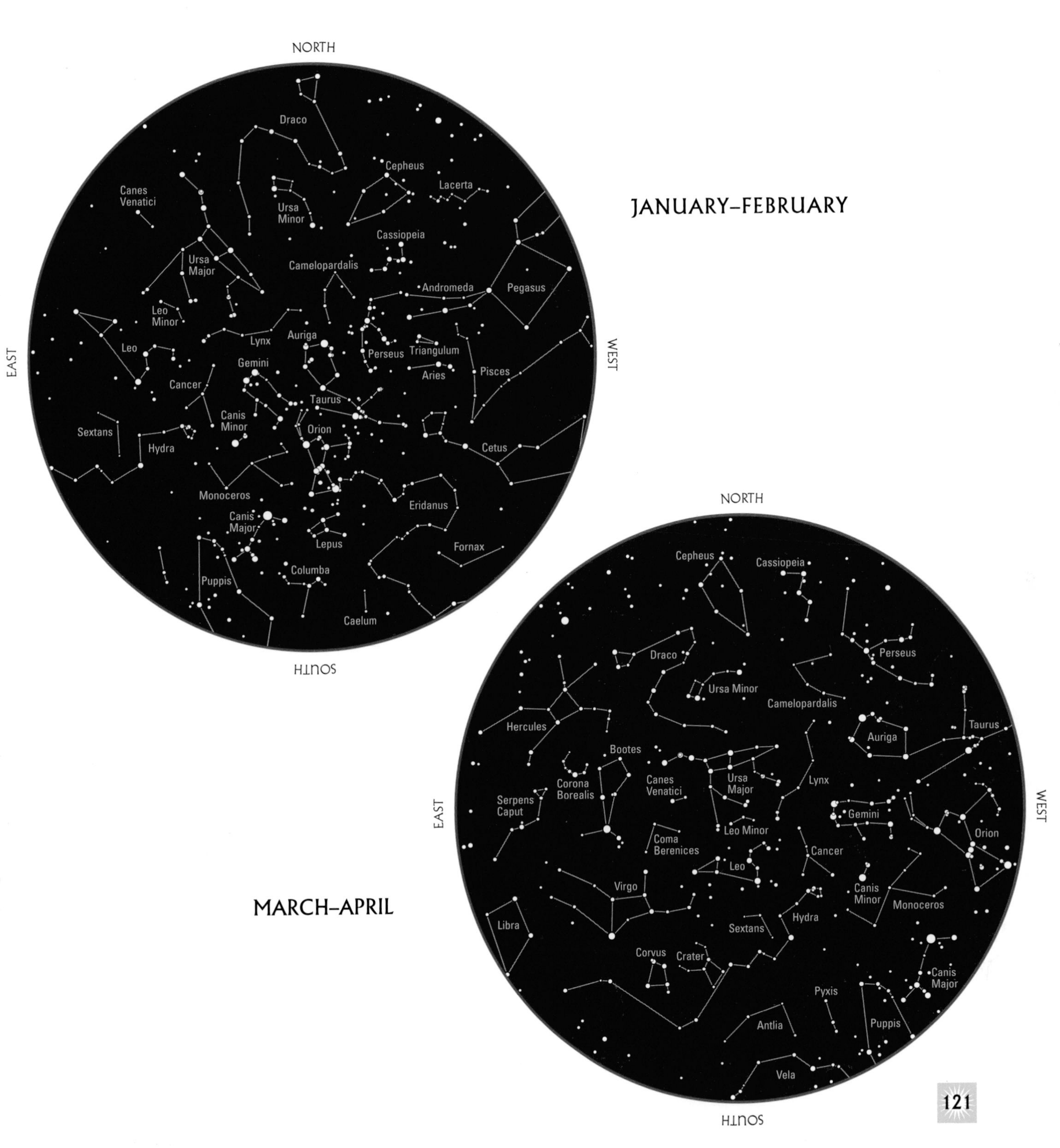

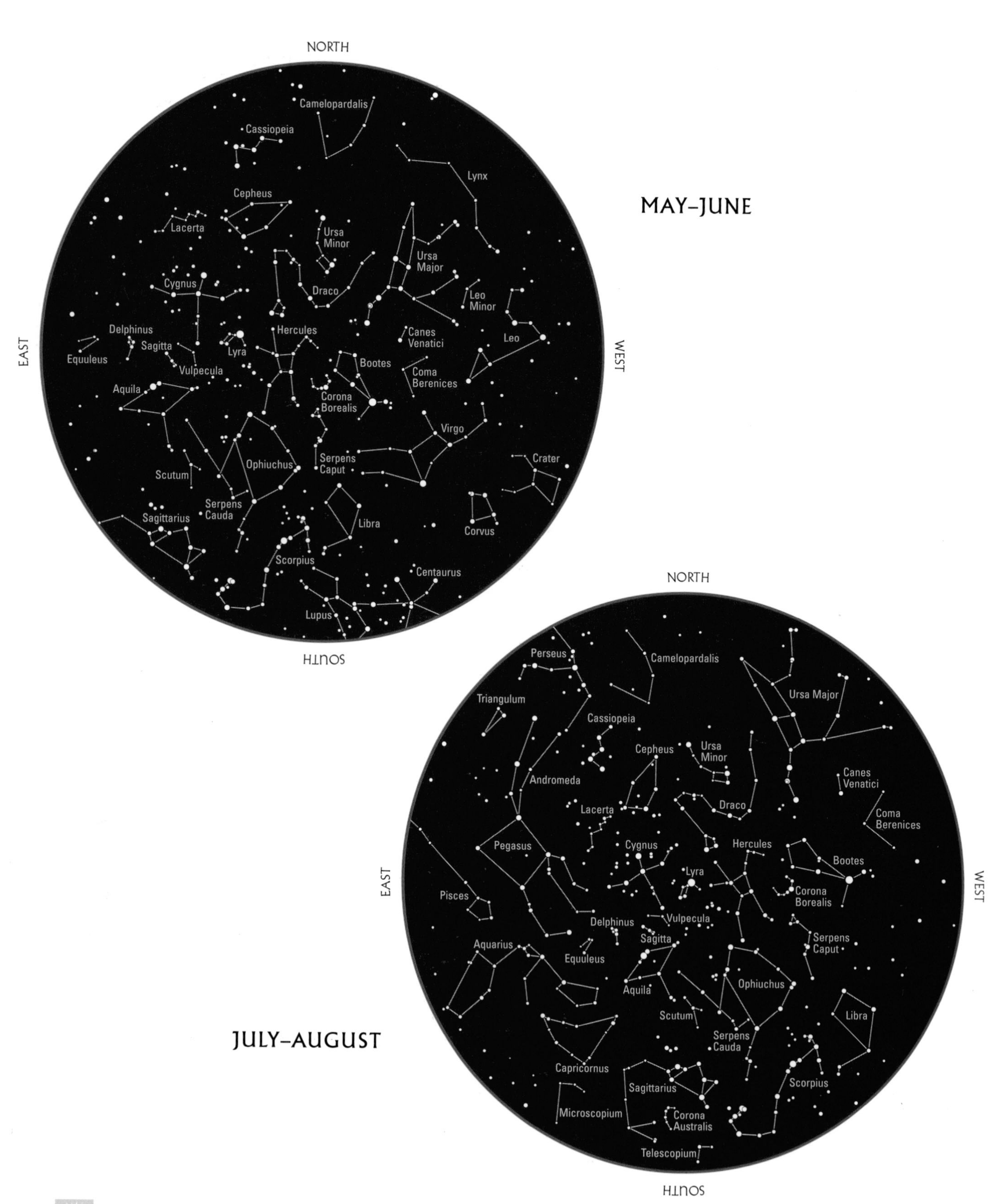
NORTH
MAY–JUNE
Camelopardalis
Cassiopeia
Lynx
Cepheus
Lacerta
Ursa Minor
Ursa Major
Cygnus
Draco
Leo Minor
Delphinus
Hercules
Canes Venatici
Leo
EAST
Sagitta
Lyra
WEST
Equuleus
Vulpecula
Bootes
Coma Berenices
Aquila
Corona Borealis
Virgo
Ophiuchus
Serpens Caput
Crater
Scutum
Serpens Cauda
Sagittarius
Libra
Corvus
Scorpius
Centaurus
Lupus
SOUTH
NORTH
JULY–AUGUST
Perseus
Camelopardalis
Triangulum
Ursa Major
Cassiopeia
Cepheus
Ursa Minor
Andromeda
Canes Venatici
Lacerta
Draco
Coma Berenices
Pegasus
Cygnus
Hercules
Bootes
Lyra
EAST
Pisces
Corona Borealis
WEST
Delphinus
Vulpecula
Sagitta
Serpens Caput
Aquarius
Equuleus
Aquila
Ophiuchus
Scutum
Libra
Serpens Cauda
Capricornus
Sagittarius
Scorpius
Microscopium
Corona Australis
Telescopium
SOUTH

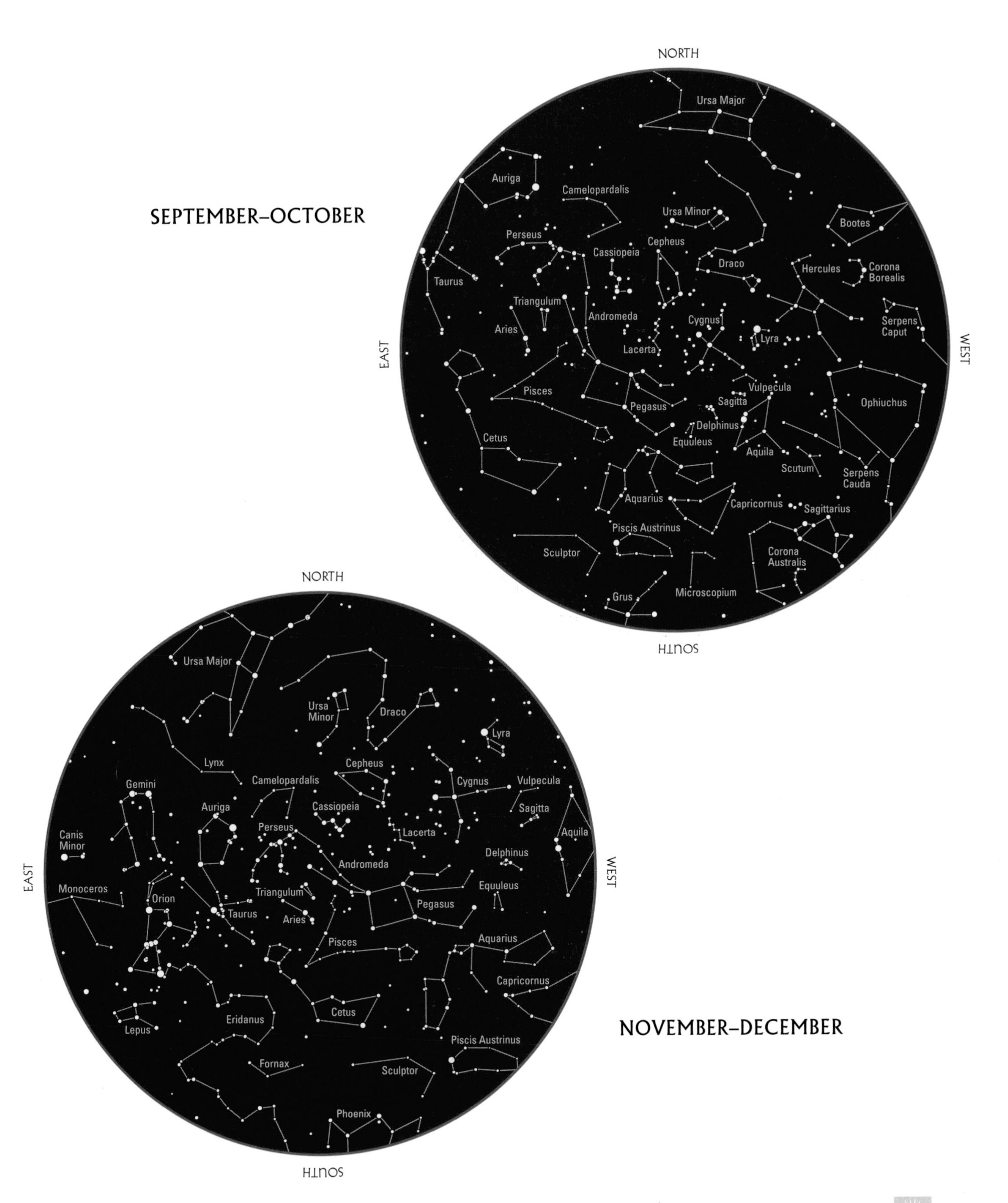
SEPTEMBER–OCTOBER
NORTH
EAST
WEST
SOUTH
Ursa Major
Auriga
Camelopardalis
Ursa Minor
Bootes
Perseus
Cepheus
Cassiopeia
Draco
Hercules
Corona Borealis
Taurus
Triangulum
Andromeda
Cygnus
Serpens Caput
Aries
Lyra
Lacerta
Pisces
Vulpecula
Ophiuchus
Sagitta
Pegasus
Delphinus
Cetus
Equuleus
Aquila
Scutum
Serpens Cauda
Aquarius
Capricornus
Sagittarius
Piscis Austrinus
Sculptor
Corona Australis
Grus
Microscopium
NOVEMBER–DECEMBER
NORTH
EAST
WEST
SOUTH
Ursa Major
Ursa Minor
Draco
Lyra
Lynx
Cepheus
Gemini
Camelopardalis
Cygnus
Vulpecula
Auriga
Cassiopeia
Sagitta
Canis Minor
Perseus
Lacerta
Aquila
Andromeda
Delphinus
Monoceros
Triangulum
Equuleus
Orion
Taurus
Pegasus
Aries
Pisces
Aquarius
Capricornus
Cetus
Eridanus
Lepus
Piscis Austrinus
Fornax
Sculptor
Phoenix

SEASONAL MAPS FOR THE SOUTHERN LATITUDES

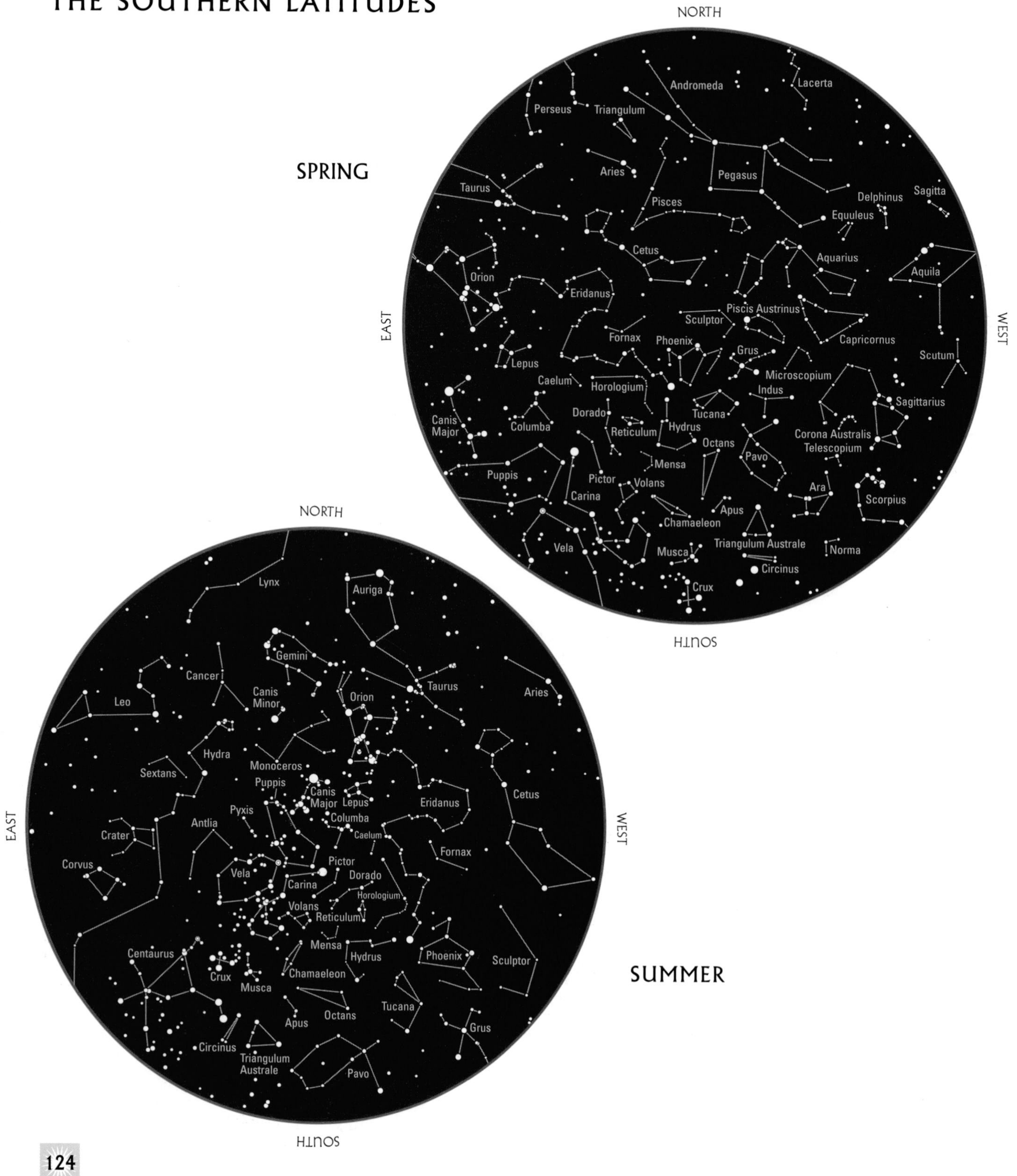

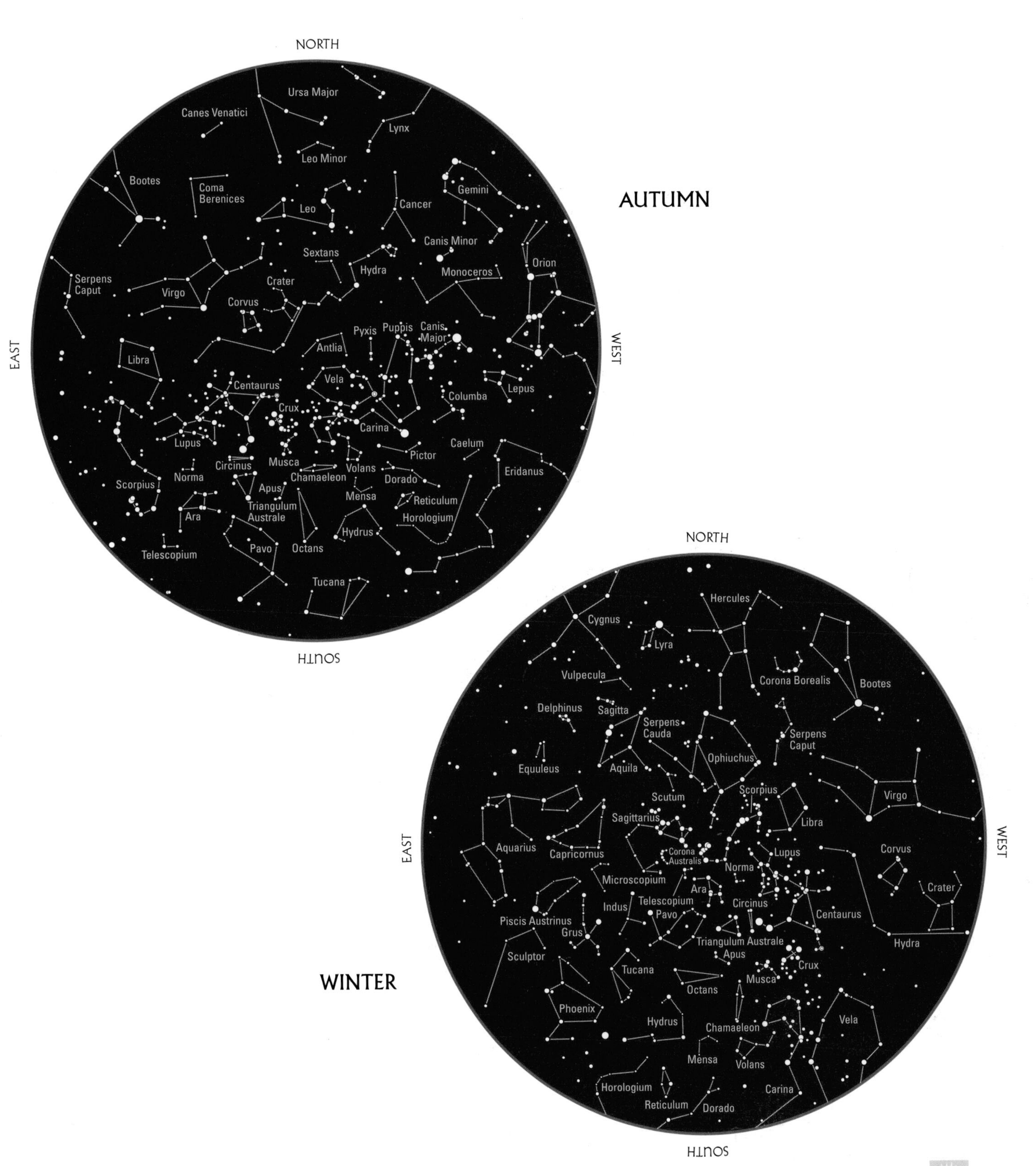
AUTUMN
NORTH
EAST
WEST
SOUTH
Ursa Major
Canes Venatici
Lynx
Leo Minor
Bootes
Coma Berenices
Gemini
Leo
Cancer
Canis Minor
Sextans
Serpens Caput
Hydra
Monoceros
Orion
Crater
Virgo
Corvus
Pyxis
Puppis
Canis Major
Antlia
Libra
Vela
Centaurus
Lepus
Columba
Crux
Carina
Lupus
Caelum
Pictor
Circinus
Musca
Norma
Volans
Eridanus
Chamaeleon
Dorado
Scorpius
Apus
Mensa
Reticulum
Triangulum Australe
Ara
Horologium
Hydrus
Pavo
Octans
Telescopium
Tucana
WINTER
NORTH
EAST
WEST
SOUTH
Hercules
Cygnus
Lyra
Corona Borealis
Vulpecula
Bootes
Delphinus
Sagitta
Serpens Cauda
Serpens Caput
Ophiuchus
Equuleus
Aquila
Scutum
Scorpius
Virgo
Sagittarius
Libra
Aquarius
Capricornus
Corona Australis
Lupus
Corvus
Norma
Microscopium
Ara
Crater
Indus
Telescopium
Circinus
Pavo
Centaurus
Piscis Austrinus
Grus
Triangulum Australe
Hydra
Apus
Sculptor
Crux
Tucana
Musca
Octans
Phoenix
Vela
Hydrus
Chamaeleon
Mensa
Volans
Horologium
Carina
Reticulum
Dorado

GLOSSARY

Asterism A recognized pattern of stars that isn't a constellation, but still has a name.

Aurora Slowly moving displays of green or pink light in the sky. The Aurora Borealis are the Northern Lights, and the Aurora Australis are the Southern Lights.

Binoculars Great for beginning stargazers. Less expensive and easier to use than a telescope. But, be careful — never look at the Sun with binoculars!

Black Hole A star that doesn't shine. A black hole's gravity is so strong, nothing can escape it—not even light!

Celestial Poles The points in the sky directly over the Earth's North and South Poles.

Comet In the night sky, comets look like faint smudges of light, sometimes with a tail. They're large pieces of ice that orbit the Sun and give off clouds of gas and dust.

Constellation One of 88 patterns of stars in the night sky. Some of the constellations on star maps were given names thousands of years ago.

Galaxy Great collections of millions or billions of stars, thousands of light-years in diameter. Our Sun is just one of hundreds of billions of stars in the Milky Way Galaxy.

Great Square A large, square-shaped asterism in Pegasus.

Hyades A cluster of stars forming the face of Taurus the Bull.

Light Pollution Light from cities that makes it hard to see the stars.

Light-year The distance light travels in one year—about 6,000,000,000,000 miles.

Local Group of Galaxies A cluster of galaxies including the Milky Way Galaxy, the Large and Small Magellanic Clouds, and the Andromeda Galaxy.

Magellanic Clouds The two galaxies closest to our own. In the southern night sky, they look like pieces of the Milky Way.

Magnitude The brightness of stars, planets, and other objects in the sky. The brighter the star, the lower the magnitude. Vega, magnitude 0, outshines Spica, magnitude 1.

Meteor A streak of light from a particle of rock falling into the Earth's atmosphere. Fireballs are very bright meteors caused by large pieces of rock or metal.

Meteor Shower Displays of meteors that happen near the same date every year. A shower gets its name from the constellation the meteors appear to radiate from.

Milky Way A faint band of light across the night sky. The Milky Way is made of the light of the billions of stars of the Milky Way Galaxy.

Moon The Earth's natural satellite. As beautiful as it is, a bright moon can make it hard to see the stars.

Nebula A great cloud of gas and dust in space. The easiest nebula to see is the Orion Nebula, hanging from the belt of Orion.

Planet A world orbiting a star. You can see five planets without the help of binoculars: Mercury, Venus, Mars, Jupiter, and Saturn.

Pleiades A beautiful cluster of stars on the shoulder of Taurus. Also called the Seven Sisters. Because of its shape, the Pleiades are often mistaken for one of the Dippers.

Proper Motion The movement of stars through the night sky. Stars move through space, but because they're very far away, only astronomers can detect this motion.

Pulsar A pulsating star. Stars smaller than the Sun can sometimes spin very rapidly. If those stars have "hot spots" on their surfaces, they blink like stellar lighthouses.

Star A great sphere of hot gas. Many stars are much bigger than the Sun—some are as wide as the orbit of Mars! And many are much smaller.

Star Cluster A group of stars held together by gravity. Open clusters don't have any particular shape. Globular clusters have thousands of stars and are shaped like spheres.

Summer Triangle A large triangular asterism in northern summer skies, made up of Vega, Altair, and Deneb.

Sun The star at the center of our solar system — an average star, except that it is the only one we know of in the entire universe (so far) with a planet that contains life.

Supernova An exploding star. When massive stars run out of fuel, they explode and can become billions of times brighter than before. A nova isn't as bright as a supernova.

Telescope An instrument that magnifies and helps us see very dim objects in the sky. *Never* look at the Sun with a telescope unless you use a *proper* solar filter—one that fits on the *front* of the telescope, not on the eyepiece closest to your eye.

Variable Star A star that changes brightness. Cepheid variable stars help astronomers tell distances in space.

Winter Hexagon A large northern winter asterism made up of the stars Castor, Pollux, Procyon, Sirius, Rigel, Aldebaran, and Capella.

Zodiac The twelve constellations that the Sun and planets move through.

INDEX